LONDON MATHEMATICAL SOCIETY STUDI

Managing editor: Professor E.B. Davies, Departme
King's College, Strand, London WC2R 2LS

1 Introduction to combinators and λ-calculus, J.R. HINDLEY & J.P. SELDIN
2 Building models by games, WILFRID HODGES
3 Local fields, J.W.S. CASSELS
4 An introduction to twistor theory, S.A. HUGGETT & K.P. TOD
5 Introduction to general relativity, L. HUGHSTON & K.P. TOD
6 Lectures on stochastic analysis: diffusion theory, DANIEL W. STROOCK
7 Dynamical systems and the theory of evolution, J. HOFBAUER & K. SIGMUND
8 Summing and nuclear norms in Banach space theory, G.J.O. JAMESON
9 Automorphisms of surfaces after Nielsen and Thurston, A. CASSON & S. BLEILER

London Mathematical Society Student Texts. 8

Summing and Nuclear Norms in Banach Space Theory

G.J.O. JAMESON

Department of Mathematics, University of Lancaster

The right of the University of Cambridge to print and sell all manner of books was granted by Henry VIII in 1534. The University has printed and published continuously since 1584.

CAMBRIDGE UNIVERSITY PRESS

Cambridge

New York New Rochelle Melbourne Sydney

CAMBRIDGE UNIVERSITY PRESS
Cambridge, New York, Melbourne, Madrid, Cape Town, Singapore, São Paulo, Delhi

Cambridge University Press
The Edinburgh Building, Cambridge CB2 8RU, UK

Published in the United States of America by Cambridge University Press, New York

www.cambridge.org
Information on this title: www.cambridge.org/9780521341349

First published 1987
Re-issued in this digitally printed version 2008

A catalogue record for this publication is available from the British Library

ISBN 978-0-521-34134-9 hardback
ISBN 978-0-521-34937-6 paperback

To the people of Nicaragua

CONTENTS

INTRODUCTION

The summing and nuclear norms of linear operators merit recognition as very basic concepts in Banach space theory, even at quite an elementary level. They have powerful applications to a variety of Banach space questions, and they generate a theory that is interesting and elegant in its own right. It is hoped that the pages that follow will go some way towards justifying these assertions. The only prerequisite is a beginner's course on normed linear spaces. As well as the confirmed Banach space specialist, our topic has something to offer to analysts whose main interest is, for example, approximation theory or operator theory.

The origins of the subject can be traced to Khinchin's inequality (published in 1923) and to Orlicz's deduction (1933) that for every unconditionally convergent series Σx_n in L_p (where $1 \leqslant p \leqslant 2$), $\Sigma \|x_n\|^2$ is convergent. In 1947, Macphail showed that in ℓ_1, such a series may have $\Sigma \|x_n\|$ divergent. Dvoretzky and Rogers then proved that the same applies in every infinite-dimensional Banach space. From this, it was a short step to define an "absolutely summing operator" to be one for which $\Sigma \|Tx_n\|$ is convergent for every unconditionally convergent series Σx_n. Further, Macphail's work showed how this property is equivalent to a certain numerical quantity being finite: this is the "1-summing norm" $\pi_1(T)$. The idea generalizes easily to give norms π_p for each finite $p \geqslant 1$. The most interesting, and "natural", cases are $p = 1,2$, and in this book our account will be largely concentrated on these cases. For operators between Hilbert spaces, π_2 coincides with the classical Hilbert-Schmidt norm.

The "nuclear" norms ν_p (for $1 \leqslant p \leqslant \infty$) are dual, in a very natural sense, to the summing norms: $\nu_{p'}$ is dual to π_p, and ν_1 to ordinary operator norm.

It is the norms themselves, rather than the corresponding classes of p-summing and p-nuclear operators, that have proved to be of such value in

Banach space theory: hence our title. In the case of ν_p, it will be enough for our purposes to confine attention to operators of finite rank. There are close connections between these norms and other numerical quantities that are the essential tools of the subject, such as projection constants, Banach-Mazur distances and basis constants. Indeed, for finite-dimensional spaces, the projection constant $\lambda(X)$ is precisely $\nu_\infty(I_X)$. The quantity $\pi_1(I_X)$ can itself be regarded as a constant characteristic of the space. It is so closely related to the projection constant that at times both can be evaluated together. We show how this calculation can be done for the spaces ℓ_1^n and ℓ_2^n.

The central theorem of the subject is Pietsch's theorem on the existence of a dominating functional for p-summing operators. In the case p = 2, this theorem combines with the nice properties of Hilbert spaces to show that (i) every 2-summing operator can be factorized through a Hilbert space, and (ii) 2-summing operators can be extended, with the value of π_2 preserved (in other words, π_2 is the notion that delivers a "Hahn-Banach" theorem for operators).

Pietsch's theorem also provides a beautifully simple proof that the projection constant of an n-dimensional space, and the distance to ℓ_2^n, are not greater than $\sqrt{n}$. This application on its own is perhaps enough to justify the claim that these norms have a rightful place is Banach space theory. Another good application is the Gordon-Lewis proof that the ordinary space of operators on ℓ_2^n has a basis constant that grows with n.

Some of the deepest results in the theory involve the comparison of different summing norms for operators between certain spaces. Theorems of this sort were initiated by Grothendieck, and the most important one is the result known as "Grothendieck's inequality", which is coming to be recognized as one of the really major theorems in Banach space theory. Part of the fascination of this theorem is its abundance of equivalent formulations. It can be (and often is) stated in terms of bilinear forms or tensor products instead of summing norms, and it has important applications in harmonic analysis. This illustrates again how intricately these norms are connected with other topics of established interest. The notions of type 2 and cotype 2 constants are the key to a wider formulation of results of this sort. In particular, the essence of Grothendieck's inequality is generalized by Maurey's theorem stating that all spaces of cotype 2 are "2-dominated".

There is a constant interplay between finite-dimensional and infinite-dimensional spaces. Some of the results are specifically concerned with finite dimensions. Others, including Grothendieck's inequality, apply to infinite-dimensional spaces, but with all the real work taking place in a finite-dimensional context. Pietsch's theorem itself does not require any element of finite dimensionality, but (as remarks above show) many of its applications do.

The summing and nuclear norms are examples (arguably the most important ones) of "operator ideal" norms, and thereby provide an introduction to the rapidly growing research area that is becoming known as operator ideals.

Concepts and definitions are introduced gently, with plenty of simple examples (these seem to be almost entirely lacking in the existing literature). Proofs are generally complete, though the details of some of the examples are left to the reader. The author is strongly committed to the principle that proofs should be as simple and direct as possible, and that they should give a "feel" for why a result is true, as well as establishing it formally. A number of the results in this book appear with a proof that is substantially simpler than the original one - though in most cases the author does not claim any of the credit for this. In other instances - such as the derivation of Khinchin's inequality with the best constant - a satisfactorily simple proof is still awaited. There are several instances where two alternative proofs are given, since both contribute something to the understanding. Results and examples are numbered consecutively in each section. Moderately important results are designated "proposition", and the most important ones "theorem". Exercises are scattered through the text, appearing at the point where they are most relevant.

The list of references is intended both to point the way to further reading and to pay some respect to those who have developed the highly satisfying theory presented here. I have endeavoured to give just enough attributions to identify the main landmarks in this development - but this does not amount to an attempt to give a systematic historical survey.

Sections 1 to 11 contain the core material, and need to be read more or less in order. The remaining sections deal with a selection of further topics, and are independent of each other.

0. BANACH SPACE BACKGROUND

Normed linear spaces, Hilbert spaces

We assume that the reader is familiar with the notions normed linear space, Banach space, inner product space, Hilbert space, and with the really basic facts about such spaces. Here we give a brief summary of the results that are particularly relevant to our purposes. Proofs are given only when it cannot be confidently asserted that they are to be found in any elementary text on the subject. At the same time, we establish some notation.

We use the same notation $\| \; \|$ for the norms in the various spaces considered, except when it is necessary to distinguish different norms. The (closed) unit ball in a normed linear space X (denoted by U_X) is the set $\{x \in X : \|x\| \leqslant 1\}$.

The scalar field may be either $\mathbb{R}$ or $\mathbb{C}$. Most results will apply to both cases simultaneously, or with minor modifications for the complex case. Exceptions to this will be pointed out.

0.1. Every Hilbert space has a (finite or infinite) orthonormal basis (b_j). For each element x,

$$x = \sum_j \langle x,b_j \rangle \, b_j \, ,$$

$$\|x\|^2 = \sum_j |\langle x,b_j \rangle|^2 \, .$$

(This means in the sense of "summation" when (b_j) is uncountable, but our main interest is in the finite-dimensional case.)

Operators

A linear operator T (from one normed linear space to another) is continuous if and only if there exists M such that $\|Tx\| \leqslant M\|x\|$ for all $x \in X$.

We will use the word "operator" to mean "continuous linear operator". The (linear) space of operators from X to Y will be denoted by L(X,Y), and we write L(X) for L(X,X). "Operator norm" is defined on L(X,Y) by:

$$\|T\| = \sup\{ \|Tx\| : x \in U_X\} .$$

This is a norm, and $\|TS\| \leq \|T\|.\|S\|$.

An operator T is an <u>isometry</u> if $\|Tx\| = \|x\|$ for all x. Spaces X,Y are said to be <u>isometric</u> if there is an isometry of X onto Y. An operator T is an <u>isomorphism</u> if it is bijective and T, T^{-1} are both continuous. Spaces X,Y are said to be <u>isomorphic</u> if there is an isomorphism of X onto Y. The <u>Banach-Mazur distance</u> between X and Y is then defined to be

$$d(X,Y) = \inf \{ \|T\|.\|T^{-1}\| : T \text{ an isomorphism of } X \text{ onto } Y\}.$$

Though not truly a "distance" (or metric), this is a measure of the similarity of X and Y. Clearly, $d(X,Y) \geq 1$, with equality when X is isometric to Y. Also, $d(X,Z) \leq d(X,Y)\, d(Y,Z)$.

We say that T is an <u>M-open</u> operator of X onto Y if, given $y \in Y$, there exists $x \in X$ with $Tx = y$ and $\|x\| \leq M\|y\|$.

Duality and the Hahn-Banach theorem

Operators mapping into the scalar field ($\mathbb{R}$ or $\mathbb{C}$) are called <u>linear functionals</u>. The space of all continuous linear functionals on X, with operator norm, is the <u>dual</u> space X^*.

The Hahn-Banach theorem is the basic theorem on extension (and existence) of linear functionals. There are two versions, as follows. A real-valued function p (on a real linear space X) is <u>sublinear</u> if $p(\lambda x) = \lambda p(x)$ and $p(x+y) \leq p(x) + p(y)$ for all $x,y \in X$ and $\lambda \geq 0$.

<u>0.2 Theorem.</u> (i) Let X be a real linear space, X_1 a linear subspace and p a sublinear real function on X. Let f_1 be a linear functional defined on X_1, with $f_1(x) \leq p(x)$ for all $x \in E$. Then there is a linear functional f on X that extends f_1 and satisfies $f(x) \leq p(x)$ for all $x \in X$.

(ii) Let X be a normed linear space (real or complex), X_1 a linear subspace. Let f_1 be a continuous linear functional defined on X_1. Then there is a linear functional f on X that extends f_1 and satisfies $\|f\| = \|f_1\|$.

0.3 Corollary. Let x_0 be an element of a normed linear space X. Then there is an element f of U_{X^*} such that $f(x_0) = \|x_0\|$.

0.4 Corollary. Any normed linear space X embeds isometrically into its second dual X^{**}, under the mapping J defined by : $(Jx)(f) = f(x)$ for $f \in X^*$. (If J maps onto X^{**}, then X is said to be reflexive).

A norming subset of U_{X^*} is a subset K such that $\|x\| = \sup\{|f(x)| : f \in K\}$ for all $x \in X$.

For T in L(X,Y), the adjoint (or dual) operator T^* in $L(Y^*,X^*)$ is defined by : $(T^*g)(x) = g(Tx)$ for $g \in Y^*$, $x \in X$.

0.5. $\|T^*\| = \|T\|$. Hence if X is isomorphic to Y, then $d(X^*,Y^*) \leq d(X,Y)$, with equality if X,Y are reflexive.

Some particular spaces

We denote by $\ell_\infty(S)$ the set of all bounded functions (real or complex, according to context) on a set S, with norm defined by : $\|x\| = \sup\{ |x(s)| : s \in S\}$. We write simply ℓ_∞ for $\ell_\infty(\mathbb{N})$, and ℓ_∞^n for $\ell_\infty(S)$ when S $= \{1,2, \dots ,n\}$; of course, ℓ_∞^n is simply $\mathbb{R}^n$ or $\mathbb{C}^n$ with the above norm. (We shall normally regard elements of $\mathbb{R}^n$, $\mathbb{C}^n$ as functions on $\{1,2, \dots ,n\}$, hence we use the notation x(j) for the jth term).

When K is a compact topological space, we denote by C(K) the space of all continuous real (or complex) functions on K, with norm defined as for $\ell_\infty(S)$.

The symbol ℓ_p^n (for $p \geq 1$) denotes $\mathbb{R}^n$ (or $\mathbb{C}^n$) with norm :

$$\|x\|_p = \Big(\sum_1 |x(i)|^p\Big)^{1/p} .$$

In particular, $\|x\|_1 = \sum_1 |x(i)|$. Further, ℓ_p denotes the space of all infinite sequences x for which $\|x\|_p$ (defined in the same way) is finite. We shall distinguish "real ℓ_p^n" and "complex ℓ_p^n" when it matters.

The norm of ℓ_2^n (or ℓ_2) is derived from the "natural" inner product:

$$\langle x, y \rangle = \sum_1 x(i)\,\overline{y(i)} .$$

Every n-dimensional Hilbert space is isometric to ℓ_2^n.

We use the notation e_j for the sequence (finite or infinite) having 1 in place j and 0 elsewhere.

The identity in $\mathbb{R}^n$ (or $\mathbb{C}^n$), regarded as an operator from ℓ_p^n to ℓ_q^n, will be denoted by $I_{p,q}^{(n)}$.

<u>0.6.</u> $\|I_{1,2}^{(n)}\| = \|I_{2,\infty}^{(n)}\| = \sqrt{n}$.

Proof. Easy, except $\|x\|_1 \leqslant \sqrt{n}\ \|x\|_2$. This follows from $\Sigma(|x(i)| - c)^2 \geqslant 0$, with $c = \frac{1}{n}\|x\|_1$.

<u>0.7.</u> The duals of ℓ_1^n, ℓ_2^n, ℓ_∞^n are isometric to ℓ_∞^n, ℓ_2^n, ℓ_1^n respectively (and the dual of ℓ_p^n to $\ell_{p'}^n$, where $\frac{1}{p} + \frac{1}{p'} = 1$). In each case, the functional corresponding to an element y is f_y, where $f_y(x) = \sum_1 x(i)y(i)$.

The spaces $L_p(\mu)$ will occasionally be mentioned in examples, but nothing of importance in this book depends on measure theory.

Finite-dimensional spaces

<u>0.8 Theorem.</u> Every linear mapping defined on a finite-dimensional normed linear space is continuous. Consequently all n-dimensional normed linear spaces (over the same field) are isomorphic.

<u>0.9 Corollary.</u> If X is finite-dimensional, then U_X is compact.

If dim X = n, then by elementary algebra, dim X* = n. Hence X is isometric to X** and d(X*,Y*) = d(X,Y). In fact, if $\{b_1, \ldots, b_n\}$ is a basis of X, then the <u>dual basis</u> of X* is $\{f_1, \ldots, f_n\}$, where the f_i are defined by : $f_i(x_j) = \delta_{ij}$. Clearly if $\|b_i\| = 1$, then $\|f_i\| \geqslant 1$.

<u>0.10 Theorem.</u> Let X be a n-dimensional normed linear space. Then there exists a basis $\{b_1, \ldots, b_n\}$ of X, with dual basis $\{f_1, \ldots, f_n\}$, such that $\|b_i\| = \|f_i\| = 1$ for all i. (Such a basis is called an <u>Auerbach basis</u>).

Proof. Take any basis $\{a_1, \ldots, a_n\}$ of X, and let T be the corresponding isomorphism of X onto $\mathbb{R}^n$ (or $\mathbb{C}^n$). Given elements $x_1, \ldots, x_n$ of X, let $D(x_1, \ldots, x_n)$ be the determinant of the matrix with columns $Tx_1, \ldots, Tx_n$. Then D is a continuous function on X^n, since it is formed by taking sums and products of coordinate functionals. Hence D attains its maximum absolute value on the compact set $(U_X)^n$, say at $(b_1, \dots, b_n)$. Write $D(b_1, \dots, b_n) = \mu$, and define

$$f_i(x) = \frac{1}{\mu} D(b_1, \dots, x, \dots, b_n)$$

(in which b_i is replaced by x). Then $f_i(b_i) = 1$, and $|f_i(x)| \leq 1$ for $x \in U_X$. By the elementary properties of determinants, f_i is linear and $f_i(b_j) = 0$ for $i \neq j$.

0.11 Corollary. If dim X = n, then $d(X, \ell_\infty^n) \leq n$, $d(X, \ell_2^n) \leq n$. If dim X = dim Y = n, then $d(X,Y) \leq n^2$.

Proof. With $\{b_j\}$ as in 0.10, let $Tx = \Sigma f_i(x)e_i \in \ell_\infty^n$. We have $\max |f_i(x)| \leq \|x\| \leq \Sigma|f_i(x)|$, hence $\|Tx\| \leq \|x\| \leq n\|Tx\|$.

Some statements about infinite-dimensional spaces are really statements about their finite-dimensional subspaces. This motivates the following definition. If X,Y are normed linear spaces, we say that Y is finitely represented in X if, given any finite-dimensional subspace Y_1 of Y and $\varepsilon > 0$, there is a subspace X_1 of X such that $d(X_1, Y_1) < 1 + \varepsilon$. This says that all the finite-dimensional subspaces of Y are "nearly isometric" to subspaces of X.

Embedding in $\ell_\infty(S)$

0.12 Proposition. Every normed linear space is isometric to a subspace of $\ell_\infty(S)$ for some set S, and to a subspace of C(K) for some compact space K.

Proof. Let $S = U_{X^*}$. Given x in X, define Jx in $\ell_\infty(S)$ by : $(Jx)(f) = f(x)$. It follows from 0.3 that $\|Jx\| = \|x\|$. The same construction proves the second statement, since U_{X^*} is compact in the weak-star topology, and Jx is continuous with respect to this topology. (Familiarity with the weak-star topology is not really needed for the purposes of this book).

For the first statement in 0.12, it is clearly enough for S to be a norming subset of U_{X^*}.

An important variation of this for finite-dimensional spaces is :

0.13 Proposition. Let $\dim X = n$ and $\varepsilon > 0$. Then there exist N and a subspace X_0 of ℓ_∞^N such that $d(X, X_0) \leqslant 1 + \varepsilon$.

Proof. The set $S_{X^*} = \{f \in X^* : \|f\| = 1\}$ is totally bounded, so contains elements $f_1, \ldots, f_N$ such that, given any $f \in S_{X^*}$, we have $\|f - f_i\| \leqslant \varepsilon$ for some i. For $x \in X$, let Jx be the element $[f_1(x), \ldots, f_N(x)]$ of ℓ_∞^N. Using 0.3, we have

$$(1 - \varepsilon)\,\|x\| \leqslant \|Jx\| \leqslant \|x\| .$$

A variant of this gives precise isometric embedding into a close copy of ℓ_∞^N.

0.14. Let $\dim X = n$ and $\varepsilon > 0$. Then there exist N and a space Y such that $d(Y, \ell_\infty^N) \leqslant 1 + \varepsilon$ and X is isometric to a subspace of Y.

Proof. Let $T : X \to \ell_\infty^N$ be such that $\|x\| \leqslant \|Tx\| \leqslant (1+\varepsilon)\|x\|$ for all x in X. Let U_∞ be the unit ball of ℓ_∞^N, and let Y be $\mathbb{R}^N$ with the norm defined by taking as unit ball the convex cover of $U_\infty \cup T(U_X)$. One verifies easily that the conditions hold.

Extensions and projections

Let X be a subspace of a normed linear space Y. A projection of Y onto X is an operator $P : Y \to X$ such that $Px = x$ for all $x \in X$. If there is such a projection, then X is said to be complemented in Y; it must then be a closed subspace, since $X = \ker(I - P)$.

0.15 Proposition. If Y is a Hilbert space, X a closed subspace, then there is a projection (the "orthogonal" projection) P of Y onto X with $\|P\| = 1$. The kernel of P is $X^\perp$. If X is finite-dimensional, then P is given by:

$$Py = \sum_i \langle y, b_i \rangle\, b_i$$

where $\{b_i\}$ is an orthonormal basis of X.

A normed linear space X is said to be injective if there is a real number λ such that the following holds: given any normed linear space E, a subspace E_1 and T_1 in $L(E_1, X)$, there is an extension T in $L(E, X)$ with

$\|T\| \leqslant \lambda\|T_1\|$. We define $\lambda(X)$ to be the infimum of such λ.

0.16. If X is isomorphic to Y, then $\lambda(Y) \leqslant d(X,Y)\,\lambda(X)$.

Proof. Elementary.

0.17 Proposition. For any set S, $\lambda[\ell_\infty(S)] = 1$.

Proof. Let E_1 be a subspace of E, and let T_1 be an operator from E_1 to $\ell_\infty(S)$. For each $s \in S$, define $f_s \in E_1^*$ by $f_s(e) = (T_1 e_1)(s)$. Then $\|f_s\| \leqslant \|T_1\|$. By the Hahn-Banach theorem, f_s can be extended to $g_s \in E^*$ with $\|g_s\| \leqslant \|T_1\|$. For $e \in E$, define $(Te)(s) = g_s(e)$.

0.18 Corollary. If $\dim X = n$, then $\lambda(X) \leqslant d(X, \ell_\infty^n) \leqslant n$.

Proof. By 0.16, 0.17, 0.11.

0.19 Proposition. The following statements (for a given space X) are equivalent:

(i) $\lambda(X) \leqslant \lambda$,

(ii) if X is isometric to a subspace X_0 of a space Y, and $\varepsilon > 0$, then there is a projection P of Y onto X_0 with $\|P\| \leqslant (1+\varepsilon)\lambda$,

(iii) for some set S, X is isometric to a subspace X_0 of $\ell_\infty(S)$, and for every $\varepsilon > 0$, there is a projection P of $\ell_\infty(S)$ onto X_0 with $\|P\| \leqslant (1+\varepsilon)\lambda$.

Proof. (i) implies (ii). We have $\lambda(X_0) = \lambda$. The projection P is obtained by extending I_{X_0}.

(ii) implies (iii), clearly.

(iii) implies (i). Let E_1 be a subspace of E, and T_1 an element of $L(E_1, X_0)$. By 0.16, there is an extension $\bar{T}$ in $L(E, \ell_\infty(S))$ with $\|\bar{T}\| = \|T_1\|$. With P as in (iii), let $T = P\bar{T}$. Then T is in $L(E, X_0)$, extends T_1 and has $\|T\| \leqslant (1+\varepsilon)\,\lambda\|T_1\|$. Hence $\lambda(X) = \lambda(X_0) \leqslant (1+\varepsilon)\lambda$.

Because of the equivalence with (ii), $\lambda(X)$ is called the projection constant of X. We will see that in fact for n-dimensional X, both $\lambda(X)$ and $d(X, \ell_2^n)$ are not greater than $\sqrt{n}$ (compare 0.11 and 0.18). We shall also describe the evaluation of the projection constants of ℓ_1^n and ℓ_2^n.

Orderings; linear lattices

The (real) spaces $\mathbb{R}^n$, ℓ_p, $\ell_\infty(S)$, $C(S)$, $L_p(\mu)$ all have a natural partial ordering defined "pointwise" : $x \leqslant y$ means $x(s) \leqslant y(s)$ for each s.

In general, a real linear space is said to be a linear lattice (or Riesz space) if it has a partial ordering $\leqslant$ such that :

(i) if $x \leqslant y$ and $y \leqslant x$, then $x = y$;

(ii) if $x \leqslant y$, then $x + z \leqslant y + z$ for all z ;

(iii) if $x \geqslant 0$ and $\lambda \geqslant 0$, then $\lambda x \geqslant 0$;

(iv) any two elements have a supremum.

The supremum of x and -x is denoted by $|x|$. The above examples are clearly linear lattices, and $|x|$ is the function given by : $|x|(s) = |x(s)|$. Note that $|x| \leqslant y$ is equivalent to $-y \leqslant x \leqslant y$.

A norm on a linear lattice is a lattice norm if :

(i) $0 \leqslant x \leqslant y$ implies $\|x\| \leqslant \|y\|$,

and (ii) $\|\,|x|\,\| = \|x\|$ for all x.

The space is then called a normed lattice. The above examples are all normed lattices.

A linear mapping T between linear lattices is positive if $Tx \geqslant 0$ whenever $x \geqslant 0$. This definition applies in particular to linear functionals, thereby giving a partial ordering of the dual space. The functional f_y on $\mathbb{R}^n$ defined by $f_y(x) = \Sigma x(i)y(i)$ is positive if and only if $y \geqslant 0$.

The spaces ℓ_1^k, ℓ_1, $L_1(\mu)$ have the special property that $\|\Sigma x_i\| = \Sigma\|x_i\|$ for positive elements (in general, normed lattices with this property are called L-spaces).

The above terminology will be used where appropriate, but we do not assume any knowledge of the general theory of normed lattices.

$\mathcal{L}_1$ and $\mathcal{L}_\infty$ spaces

These notions are the key to the extension of certain results to the infinite-dimensional case. However, they can be omitted without serious loss.

We will say that a Banach space X is an $\mathcal{L}_p$-space if for every finite-dimensional subspace E of X and $\varepsilon > 0$, there is a finite-dimensional subspace F such that $E \subseteq F \subseteq X$ and $d(F, \ell_p^N) \leqslant 1 + \varepsilon$, where $N = \dim F$. (This is not quite the usual terminology; according to this, X is an "$\mathcal{L}_{p,\lambda}$-space" if we have $d(F,\ell_p^N) \leqslant \lambda$; hence our definition equates to an "$\mathcal{L}_{p,1+\varepsilon}$-space for every $\varepsilon > 0$".)

We only need the fact the certain naturally arising spaces are $\mathcal{L}_\infty$ or $\mathcal{L}_1$ spaces. The next lemma is useful for this purpose.

<u>0.20 Lemma.</u> The following is sufficient for X to be an $\mathcal{L}_p$-space. Given $b_1, \ldots, b_n \in X$ and $\varepsilon > 0$, there is a finite-dimensional subspace F of X such that $d(F, \ell_p^N) \leqslant 1+\varepsilon$ (where N = dim F) and $\operatorname{dist}(b_i, F) < \varepsilon$ for each i.

Proof. Given E, let $\{b_i\}$ be an Auerbach basis of E. . Let F be as stated, with ε replaced by $\varepsilon' = \varepsilon/2n^2$. For each i, take $f_i \in F$ with $\|b_i - f_i\| < \varepsilon'$. Essentially, we modify F by replacing the b_i's by the f_i's. Define $J : E \to F$ by $Jb_i = f_i$. One verifies easily that $\|b - Jb\| \leqslant n\varepsilon'\|b\|$ for $b \in B$. We may assume $n\varepsilon' \leqslant \frac{1}{2}$: then $\|b - Jb\| \leqslant 2n\varepsilon'\|Jb\|$. Let P be a projection of F onto J(B) with $\|P\| \leqslant n$ (see 0.18), and define T on F by :

$$Tx = J^{-1}Px + (I - P)x .$$

Then T(F) contains B, and

$$Tx - x = J^{-1}Px - Px = b - Jb ,$$

where $b = J^{-1}Px$. Hence $\|Tx - x\| \leqslant 2n\varepsilon'\|Px\| \leqslant 2n^2\varepsilon'\|x\| = \varepsilon\|x\|$, from which $\|T\|.\ \|T^{-1}\| \leqslant (1+\varepsilon)/(1-\varepsilon)$.

<u>0.21.</u> ℓ_1 is an $\mathcal{L}_1$-space, and c_0 (the space of sequences tending to 0) is an $\mathcal{L}_\infty$-space.

Proof. Let $E_N = \operatorname{lin}(e_1, \ldots, e_N)$. This is isometric to ℓ_1^N, ℓ_∞^N in the two cases. Given elements $b_1, \ldots, b_n$ and $\varepsilon > 0$, there exists N such that $\operatorname{dist}(b_i, E_N) < \varepsilon$ for each i.

<u>0.22.</u> If μ is a positive measure on a measure space, then $L_1(\mu)$ is an $\mathcal{L}_1$-space, $L_\infty(\mu)$ an $\mathcal{L}_\infty$-space.

Proof. In both cases, simple functions are dense. Hence if elements $b_1, \ldots, b_n$ and $\varepsilon > 0$ are given, there are disjoint measurable sets $A_1, \ldots, A_N$ (with finite measure in the case of $L_1(\mu)$) such that $\operatorname{dist}(b_i, F) \leqslant \varepsilon$ for each i, where F is the subspace spanned by the characteristic functions of the A_i. It is easily seen that F is isometric to ℓ_1^N, ℓ_∞^N in the two cases.

In particular, $\ell_\infty(S)$ is an $\mathcal{L}_\infty$-space. By 0.12, it is clear that the property of being an $\mathcal{L}_\infty$-space is not inherited by subspaces.

For readers with sufficient grounding in General Topology, we show also that C(K) is an $\mathcal{L}_\infty$-space. We use the fact that if $\{G_1, \dots, G_N\}$ is an open covering of a compact, Hausdorff space K, then there exist non-negative continuous functions g_i such that $g_1 + \dots + g_N = 1$ and $g_i(s) = 0$ for s not in G_i (a "partition of unity"). The covering might as well be chosen so that each G_i contains a point s_i not in the other G_j : clearly, we then have $g_i(s_i) = 1$.

<u>0.23 Lemma.</u> Under these conditions, $\mathrm{lin}(g_1, \dots, g_N)$ is isometric to ℓ_∞^N .

Proof. Let $g = \sum_i \lambda_i g_i$. Evaluation at s_i shows that $\|g\| \geq |\lambda_i|$ for each i. Conversely, if $M = \max |\lambda_i|$, then

$$|g(s)| \leq M \sum_i g_i(s) = M$$

for all s.

<u>0.24 Proposition.</u> If K is a compact, Hausdorff space, then C(K) is an $\mathcal{L}_\infty$-space.

Proof. Take elements $f_1, \dots, f_n$ and $\varepsilon > 0$. Let M be an integer such that $\|f_i\| \leq M\varepsilon$ for each i, and write the numbers $r\varepsilon$ $(r = -M, \dots, M-1, M)$ as $\lambda_1, \dots, \lambda_k$. For $(r_1, \dots, r_n)$ in $\{1, \dots, k\}^n$, let

$$G(r_1, \dots, r_n) = \{s : |f_i(x) - \lambda_{r_i}| < \varepsilon \text{ for each } i\} .$$

These sets form an open covering: remove any that are empty or contained in the others, and let B be the set of remaining $(r_1, \dots, r_n)$. Choose corresponding functions $g_{r_1, \dots, r_n}$ to form a partition of unity. We must show that each f_i is close to the linear subspace spanned by these functions. We do this for f_1. Let

$$h_1 = \Sigma \{\lambda_{r_1}, \dots, r_n \cdot (r_1, \dots, r_n) \in B\} .$$

Choose $s \in K$. There exists p such that $\lambda_p \leq f_1(s) < \lambda_p + \varepsilon$. Then s belongs to $G(r_1, \dots, r_n)$ only for $r_1 = p$ and $r_1 = p + 1$ (in which case $\lambda_{r_1} = \lambda_p + \varepsilon$). Hence

$$h_1(s) = \lambda_p \Sigma g_{p, r_2, \dots, r_n}(s) + (\lambda_p + \varepsilon) \Sigma g_{p+1, r_2, \dots, r_n}(s)$$

(summation over all elements of B of this form)

$$= \lambda_p + \varepsilon \Sigma g_{p+1, r_2, \dots, r_n}(s) .$$

Hence $\lambda_p \leqslant h_1(s) < \lambda_{p+1}$, so $|f_1(s) - h_1(s)| \leqslant \varepsilon$.

The argument can be adapted for the complex case.

A deeper analysis (see, e.g. [CBS]) shows in fact that $\mathcal{L}_1$ and $\mathcal{L}_\infty$ spaces (in the sense defined here) can be characterized as follows:

(a) X is an $\mathcal{L}_1$-space if and only if X is isometric to $L_1(\mu)$ for some μ,

(b) X is an $\mathcal{L}_\infty$-space if and only if X* is an $\mathcal{L}_1$-space.

The following, weaker result concerning the dual of an $\mathcal{L}_\infty$-space is easier to prove and sufficient for our purposes.

0.25. If X is an $\mathcal{L}_\infty$-space, then X* is finitely represented in ℓ_1.

Proof. Let F be a finite dimensional subspace of X*, and take $\varepsilon > 0$. There is a finite-dimensional subspace E of X such that for each $f \in F$, we have $\|Rf\| \geqslant (1-\varepsilon) \|f\|$, where $Rf = f|_E$. (To prove this, take f_i as in 0.13 and $x_i \in U_X$ with $f_i(x_i) > 1-\varepsilon$). By taking a larger subspace if necessary, we may assume that $d(E, \ell_\infty^N) \leqslant 1+\varepsilon$. It follows that $d(E^*, \ell_1^N) \leqslant 1+\varepsilon$, and hence (by restriction of the isomorphism involved) that $d[R(F),G] \leqslant 1+\varepsilon$, where G is a subspace of ℓ_1^N. So $d(F,G) \leqslant (1+\varepsilon)/(1-\varepsilon)$.

1. FINITE RANK OPERATORS: TRACE AND 1-NUCLEAR NORM

Representation of finite-rank operators

Let X,Y be normed linear spaces. We denote by FL(X,Y) the space of all continuous linear operators from X to Y with finite-dimensional range. The dimension of the range is called the rank of the operator.

Given $f \in X^*$ and $y \in Y$, we denote by $f \otimes y$ the rank-one operator $T : X \to Y$ defined by : $Tx = f(x)y$ for $x \in X$. Clearly, we have $\|f \otimes y\| = \|f\| \cdot \|y\|$.

Any element T of FL(X,Y) is expressible (in many different ways) in the form

$$\sum_{i=1}^{m} f_i \otimes y_i \ , \tag{1}$$

with $m \geq$ rank T. For example, if $(b_1, \ldots, b_n)$ is a basis of the range T(X), then there are unique elements f_i of X^* such that for each $x \in X$,

$$Tx = \sum_{i=1}^{n} f_i(x) b_i \ .$$

Note that if $(b_1, \ldots, b_n)$ is a basis of X, and $(f_1, \ldots, f_n)$ is the dual basis of X^*, then

$$I_X = \sum_{1}^{n} f_i \otimes b_i \ . \tag{2}$$

Representations of the form (1) do not need to have the elements y_i belonging to T(X). For instance, if E is a subspace of X, then the expression in (2) can serve as a representation of I_E (regarded as a mapping $E \to X$), even when none of the b_i belong to E.

We start with two very simple results on representations.

1.1 Let T be in FL(X,Y), and suppose that $T = \sum_i f_i \otimes y_i$.

Then :

(i) for U in L(Y,Z), $UT = \sum_i f_i \otimes (Uy_i)$,

(ii) for S in L(V, X), $TS = \sum_i (S^*f_i) \otimes y_i$.

Proof. For $x \in X$ and $v \in V$, we have

$$UTx = \sum_i f_i(x)(Uy_i) ,$$

$$TSv = \sum_i f_i(Sv)y_i .$$

In particular, given a basis $(b_1, \ldots ,b_n)$ of a finite-dimensional space X, with dual basis $(f_1, \ldots ,f_n)$, we have "natural representations:

$$T = \sum_i f_i \otimes (Tb_i) \qquad \text{for } T \in L(X,Y),$$

$$S = \sum_i (S^*f_i) \otimes b_i \qquad \text{for } S \in L(V,X) .$$

1.2. Let T be in FL(X,Y), with $T = \sum_i f_i \otimes y_i$. Then

$$T^* = \sum_i \hat{y}_i \otimes f_i , \qquad T^{**} = \sum_i \hat{f}_i \otimes \hat{y}_i ,$$

where $y \to \hat{y}$ is the natural embedding of a space into its second dual.

Proof. We have $Tx = \sum_i f_i(x)y_i$ for $x \in X$. So for $g \in Y^*$,

$$(T^*g)(x) = g(Tx) = \sum_i f_i(x)g(y_i).$$

Hence $T^*g = \sum_i g(y_i)f_i$. This proves the statement for T^*, and the statement for T^{**} follows.

Trace

The reader may already be familiar with the following simple algebraic facts.

1.3 Proposition. Let T be FL(X), and suppose that

$$T = \sum_{i=1}^{m} f_i \otimes x_i = \sum_{j=1}^{n} g_j \otimes y_j .$$

Then $\sum_i f_i(x_i) = \sum_j g_j(y_j)$.

Proof. We show by induction that if $\sum_{i=1}^{n} f_i \otimes x_i = 0$, then $\Sigma f_i(x_i) = 0$. This is clearly true for n = 1. Assume it for a certain n, and suppose that

$$\sum_{i=1}^{n+1} f_i \otimes x_i = 0 .$$

If the x_i are linearly independent, then $f_i = 0$ for all i, so the statement holds. Assume that the x_i are linearly dependent, so that (with suitable indexing)

$$x_{n+1} = \sum_{i=1}^{n} \lambda_i x_i .$$

Then

$$\sum_{i=1}^{n} (f_i + \lambda_i f_{n+1}) \otimes x_i = 0 .$$

so the induction hypothesis gives

$$\begin{aligned} 0 &= \sum_{i=1}^{n} (f_i + \lambda_i f_{n+1})(x_i) \\ &= \sum_{i=1}^{n} f_i(x_i) + f_{n+1}(x_{n+1}) , \end{aligned}$$

as required.

Hence, for T in FL(X), we can define the <u>trace</u> of T to be $\sum_i f_i(x_i)$, where $\sum_i f_i \otimes x_i$ is any representation of T. Trace is clearly a linear functional on FL(X). Note that by 1.2, trace T^* = trace T.

<u>1.4 Example.</u> Let X have basis $(b_1, \ldots, b_n)$, and let $T \in L(X)$ be given by : $Tb_j = \sum_i \alpha_{ij} b_i$, (so (α_{ij}) is the matrix of T with respect to (b_j)). Then trace $T = \sum_j \alpha_{jj}$.

Proof. Let (f_i) be biorthogonal to (b_i). Then

$$\text{trace } T = \sum_j f_j(Tb_j) = \sum_j \alpha_{jj} .$$

<u>1.5 Example.</u> Let H be a finite-dimensional Hilbert space, with orthogonal basis $(e_1, \ldots, e_n)$. Then for T in L(H),

$$\text{trace } T = \sum_i \langle Te_i, e_i \rangle .$$

Proof. This follows from the fact that $Tx = \Sigma \langle x,e_i \rangle Te_i$ for $x \in H$.

<u>1.6</u> (i) If dim X = n, then trace $I_X = n$.

(ii) If P is a projection of rank n, then trace P = n.

Proof. (i) is a special case of (ii). Let P be a projection onto E, and let $(b_1, \ldots ,b_n)$ be a basis of E. Then

$$Px = \sum_{i=1}^{n} f_i(x)b_i \,.$$

for $x \in X$, where $f_i(b_j) = \delta_{ij}$, since $Pb_i = b_i$. Hence

$$\text{trace } P = \sum_i f_i(b_i) = n \,.$$

Trace is not defined on FL(X,Y), where $Y \neq X$. However, if S is in FL(X,Y) and T is in L(Y,X), then trace (TS) exists. In this context, we have:

<u>1.7.</u> If S is in FL(X,Y) and T is in L(Y,X), then

trace (TS) = trace (ST).

Proof. Let $S = \Sigma f_i \otimes y_i$. Then $TS = \Sigma f_i \otimes (Ty_i)$, hence trace (TS) $= \Sigma f_i(Ty_i)$. Also, by 1.1, $ST = \Sigma(T^*f_i) \otimes y_i$, so trace (ST) $= \Sigma (T^*f_i)(y_i) = \Sigma f_i(Ty_i)$.

For each T in L(Y,X), a linear functional ϕ_T on FL(X,Y) is defined by : $\phi_T(S)$ = trace (TS). Furthermore, if X is finite-dimensional, then all functionals on L(X,Y) are of this form:

<u>1.8.</u> Let X be finite-dimensional, and let ϕ be a linear functional on L(X,Y). Then there exists T in L(Y,X) such that $\phi(S)$ = trace (TS) for all S in L(X,Y).

Proof. Given any $y \in Y$, define a corresponding functional ψ on X^* by : $\psi(f) = \phi(f \otimes y)$. The functional ψ corresponds to an element Ty of X, that is : $\phi(f \otimes y) = f(Ty)$ for all f in X^*. This defines an element T of L(Y,X), and if $S = f \otimes y$, we have $\phi(S) = f(Ty)$ = trace (TS) . Hence the same holds for all S in L(X,Y).

Remarks (1) When both X and Y are finite-dimensional, elements S of L(X,Y) correspond to matrices $(\alpha_{i,j})$, and of course an element ϕ of the dual identifies with a matrix $(\beta_{j,i})$ by the relation

$$\phi(S) = \sum_i \sum_j \alpha_{i,j} \beta_{j,i} .$$

It is easy to reconcile this with 1.8 : the expression is exactly trace (TS), where T is the operator in L(Y,X) corresponding to $(\beta_{j,i})$.

(2) The proof of 1.8 remains valid (for functionals on FL(X,Y)) if X is reflexive and there is a K such that $|\phi(f \otimes y)| \leq K\|f\|.\|y\|$ for all f, y. If X is not reflexive, the same reasoning gives an operator T that maps into X** instead of X ; we return to this question in section 17.

Now let X,Y be any normed linear spaces, and let α be some norm defined at least on FL(X,Y) (not necessarily on the whole of L(X,Y)). Define

$$\alpha^*(T) = \sup \{|\text{trace } (TS)| : S \in FL(X,Y) , \alpha(S) \leq 1\}$$

for those elements T of L(Y,X) for which this is finite (denote the set of such T by $L_{\alpha^*}(Y,X)$). Then α^* is a norm on $L_{\alpha^*}(X,Y)$: it is called the dual (or conjugate) norm to α under trace duality. Of course, $\alpha^*(T)$ is simply the norm of ϕ_T as a functional on $[FL(X,Y),\alpha]$, and 1.8 shows that if X is finite-dimensional, then $L_{\alpha^*}(Y,X)$ identifies with the Banach space dual of $[L(X,Y),\alpha]$.

Note that this definition has been framed in an unsymmetrical manner : α is only applied to finite-rank operators, but α^* it not restricted in this way. The point of doing this will soon become apparent. However, when both X and Y are finite-dimensional, everything is straightforward : the dual of $[L(X,Y),\alpha]$ is $[L(Y,X),\alpha^*]$, and conversely.

Exercise. Let X be finite-dimensional, and let α be a norm on L(X,Y) such that $\alpha(S) \geq \|S\|$ for all S. Show that $\alpha^*(g \otimes x) \leq \|g\|.\|x\|$ for all $g \in Y^*$, $x \in X$, and hence that $L_{\alpha^*}(Y,X)$ is the whole of L(Y,X) .

We now proceed to identify the norm α for which α^* is ordinary operator norm.

1-nuclear norm

Let X,Y be normed linear spaces. The 1-nuclear norm ν_1 is defined on FL(X,Y) as follows:

$$\nu_1(T) = \inf \{ \sum_i \|f_i\|.\|y_i\| : T = \sum_i f_i \otimes y_i \},$$

in which all finite representations of T are considered, of whatever length. Equivalently, $\nu_1(T)$ is the infimum of $\Sigma \|T_i\|$ over expressions of T as ΣT_i, a finite sum of rank-one operators.

1.9. ν_1 is a norm, and $\nu_1(T) \geqslant \|T\|$. Further :

(i) $\nu_1(f \otimes y) = \|f\|.\|y\|$,

(ii) $\nu_1(BT) \leqslant \|B\| \nu_1(T)$, $\nu_1(TA) \leqslant \nu_1(T) \|A\|$ (whenever BT and TA are defined),

(iii) $\nu_1(T^*) \leqslant \nu_1(T)$, with equality when Y is finite-dimensional (or reflexive).

Proof. The equality $\nu_1(\lambda T) = |\lambda| \nu_1(T)$ is trivial. The fact that $\nu_1(T_1+T_2) \leqslant \nu_1(T_1) + \nu_1(T_2)$ is seen by combining suitable representations for T_1 and T_2 (note that this will give a representation of greater length).

If $T = \Sigma f_i \otimes y_i$ and $\|x\| \leqslant 1$, then

$$\|Tx\| \leqslant \Sigma |f_i(x)|.\|y_i\| \leqslant \Sigma \|f_i\|.\|y_i\|.$$

Hence $\|T\| \leqslant \Sigma \|f_i\|.\|y_i\|$, so $\|T\| \leqslant \nu_1(T)$. This completes the proof that ν_1 is a norm, and (i) follows, since $\|f \otimes y\| = \|f\|.\|y\|$.

(ii) and (iii) follow from the corresponding statements in 1.1 and 1.2 (note that if Y is reflexive, then all representations of T^* are of the form given in 1.2).

We list some further immediate comments:

(1) Clearly we have $|\text{trace } T| \leqslant \nu_1(T)$ for T in FL(X).

(2) If T_1 is a restriction of T, then $\nu_1(T_1) \leqslant \nu_1(T)$.

(3) The extension problem is trivial for ν_1. If X_1 is a subspace of X, and we are given T_1 in FL(X_1,Y), then, for any $\varepsilon > 0$, there is an extension T of T_1 in FL(X,Y) with $\nu_1(T) \leqslant (1+\varepsilon)\nu_1(T_1)$. This is obtained by choosing a suitable representation $\Sigma f_i \otimes y_i$ for T_1 and taking Hahn-Banach extensions of the f_i,

(4) Any norm (for operators) satisfying conditions (i) and (ii) of 1.10 is called an operator ideal norm. We shall be meeting many further

examples of such norms. From the way it is defined, it is clear that ν_1 is the largest possible operator ideal norm.

Exercise. Show that for T in $L(X,Y^*)$, $\nu_1(T^*) = \nu_1(T)$. (Consider the mapping $\hat{T} : Y \to X^*$ defined by : $(\hat{T}y)(x) = (Tx)(y)$.).

The Auerbach basis theorem (0.10) gives :

1.10. (i) If dim X = n, then $\nu_1(I_X) = n$.

(ii) If rank T = n, then $\nu_1(T) \leq n\|T\|$.

(iii) Let E be an n-dimensional subspace of X. Then there is a projection P of X onto E with $\nu_1(P) = n$ (and any projection has $\nu_1(P) \geq n$).

Proof. (i) An Auerbach basis expresses I_X as $\sum_1^n f_i \otimes b_i$, with $\|f_i\| = \|b_i\| = 1$ for each i. Hence $\nu_1(I_X) \leq n$. Conversely, $\nu_1(I_X) \geq$ trace $I_X = n$.

(ii) Let $T(X) = Z$. Then dim Z = n, and $T = I_Z T$, so

$$\nu_1(T) \leq \nu_1(I_Z)\|T\| = n\|T\|.$$

(iii) As in (i), $\nu_1(P) \geq$ trace $P = n$. To obtain equality, take an Auerbach basis (b_i,f_i) of E and extend the functionals f_i.

The promised result on duality is completely straightforward:

1.11 Proposition. Under trace duality, the dual of ν_1 is ordinary operator norm. In other words, for T in L(Y,X),

$$\|T\| = \sup \{ |\text{trace}\ (TS)| \ : \ S \in FL(X,Y),\ \nu_1(S) \leq 1 \}$$

and if X is finite-dimensional, then the Banach space dual of $[L(X,Y),\nu_1]$ identifies with $[L(Y,X), \|\ \|]$.

Proof. If $\nu_1(S) \leq 1$, then

$|\text{trace}\ (TS)| \leq \nu_1(TS) \leq \|T\|\ \nu_1(S) \leq \|T\|$.

Take $\varepsilon > 0$. There exists $y_0 \in Y$ with $\|y_0\| = 1$ and $\|Ty_0\| \geq (1-\varepsilon)\|T\|$. There exists $f_0 \in X^*$ with $\|f_0\| = 1$ and

$$f_0(Ty_0) = \|Ty_0\| \geq (1-\varepsilon)\|T\|.$$

Let $S_0 = f_0 \otimes y_0$. Then $\nu_1(S_0) = 1$ and $f_0(Ty_0) = \text{trace}\ (TS_0)$. The statement follows.

By the Hahn-Banach theorem, we deduce:

<u>1.12 Corollary.</u> If X is finite-dimensional and S is in L(X,Y), then there exists T in L(Y,X) with $\|T\| = 1$ and $\text{trace}\ (TS) = \nu_1(S)$.

(A direct proof of this is by no means as simple as 1.11 !).

By remark (2) after 1.8, we can allow X to be reflexive instead of finite-dimensional in 1.11 and 1.12.

The dual of $\|\ \|$, in the sense that we have defined, is not ν_1, since an operator for which $\|T\|^*$ is finite certainly does not need to be of finite rank. We return to this question (which is not really central to our purposes) in section 16.

Using 1.12, we can derive a variant of 1.9 (iii).

<u>1.13.</u> If X is finite-dimensional (or reflexive) and S is in FL(X,Y), then $\nu_1(S^*) = \nu_1(S)$.

Proof. We have already $\nu_1(S^*) \leq \nu_1(S)$. Let T be as in 1.12

Then

$$\nu_1(S) = \text{trace}\ (TS) = \text{trace}\ (S^*T^*) \leq \nu_1(S^*)$$

since $\|T^*\| = 1$.

One can in fact show that $\nu_1(S^*) = \nu_1(S)$ without restrictions on X or Y. This is best proved using "local reflexivity", and we defer it to section 17.

<u>Note.</u> An operator T is said to be "nuclear" if it is expressible as $\sum_1^\infty f_i \otimes y_i$ with $\sum_1 \|f_i\|.\|y_i\|$ finite. A norm is then defined by taking the infimum of such sums. This is the usual definition of "nuclear norm". Even for finite rank operators, it allows representations of infinite length, and consequently can differ from our ν_1. However, one can show that the two definitions do coincide when either X or Y is finite dimensional. For the purposes of this book, it will be enough to confine attention to finite-rank operators and ν_1 as we have defined it.

Some particular cases

It is not, in general, at all easy to compute $\nu_1(T)$ for a particular operator T. However, there are certain situations where it is relatively simple, and we now describe a few of them.

1.14. For any operator T on ℓ_∞^n, $\nu_1(T) = \sum_1^n \|Te_i\|$.

Proof. The obvious expression $Tx = \sum_i x(i)\, Te_i$ shows that $\nu_1(T) \leq \sum_i \|Te_i\|$.

Suppose that $T = \sum_j f_j \otimes y_j$, with $\|y_j\| = 1$ for each j. Then $\|Tx\| \leq \sum_j |f_j(x)|$ for all x. Now $\|f_j\| = \sum_i |f_j(e_i)|$, so

$$\sum_i \|Te_i\| \leq \sum_j \sum_i |f_j(e_i)| = \sum \|f_j\| .$$

This shows that $\nu_1(T) \geq \sum_i \|Te_i\|$.

1.15 Let T be the operator from X into ℓ_1^n given by : $Tx = \sum_i f_i(x)\, e_i$ (so that $f_i = T^*e_i$). Then $\nu_1(T) = \sum_i \|f_i\|$.

Proof. By 1.9, $\nu_1(T) = \nu_1(T^*)$. The statement follows, by 1.14. (A direct proof is also easy).

For operators between finite-dimensional Hilbert spaces, a characterisation of $\nu_1(T)$ can be derived from the spectral theorem. This can be stated (for both real and complex scalars) as follows: there exist finite orthonormal sequences (e_j) in X, (f_j) in Y and numbers $\lambda_j \geq 0$ such that

$$Tx = \sum_{j=1}^n \lambda_j \langle x, e_j \rangle f_j \qquad \text{for } x \in X.$$

The numbers λ_j^2 are the eigenvalues of T^*T.

1.16. Let X, Y be finite-dimensional Hilbert spaces. Let T be in L(X,Y), with spectral decomposition as above. Then

$$\nu_1(T) = \sum_j \lambda_j$$
$$= \inf \{ \sum_i \|Tb_i\| : (b_i) \text{ an orthonormal base of } X \} .$$

Proof. If (b_j) is any orthornomal base of X, then

$$Tx = \sum_j \langle x,b_j \rangle Tb_j$$

for $x \in X$, which shows that $\nu_1(T) \leqslant \sum_j \|Tb_j\|$. Also, the given expression implies that $\nu_1(T) \leqslant \Sigma\lambda_j$.

There is an operator S in L(Y,X) with $\|S\| = 1$ and $Sf_j = e_j$ for each j. Clearly, trace (ST) = $\Sigma\lambda_j$. Hence $\nu_1(T) \geqslant$ trace (ST) = $\Sigma\lambda_j$.

(Though informative, this result is not actually needed for our later developments).

Exercise. Let T be in FL(X,Y) and $\varepsilon > 0$. Show that for some k there are operators T_1 in L(X, ℓ_∞^k) and T_2 in L(ℓ_∞^k ,Y) such that $T = T_2T_1$ and $\|T_1\| = 1$, $\nu_1(T_2) \leqslant (1+\varepsilon)\,\nu_1(T)$.

Dependence on the range space

Suppose that T is in FL(X,Y), and that its range is contained in a certain subspace Y_1 of Y. Let T_1 be the same mapping regarded as an element of FL(X,Y_1). It is clear that we must distinguish T_1 from T when considering the nuclear norm. Among the representations $\Sigma\, f_i \otimes y_i$ of T, only those having $y_i \in Y_1$ are allowable for T_1. Consequently, $\nu_1(T_1) \geqslant \nu_1(T)$.

The following example illustrates the distinction nicely.

1.17 Example. Let $Y = \ell_1^n$, and

$$Y_1 = \{y \in Y \; : \; \sum_i y(i) = 0\},$$

so that $\dim Y_1 = n-1$. Let $T_1 = I_{Y_1}$, and let T be the same mapping regarded as an element of L(Y_1 ,Y) .

By 1.10, $\nu_1(T_1) = n-1$. For T, we have the obvious expression

$$T = \sum_{i=1}^{n} e_i' \otimes e_i ,$$

in which e_i' is the functional on Y_1 defined by $e_i'(y) = y(i)$. It is elementary that $\|e_i'\| = \frac{1}{2}$. Hence $\nu_1(T) \leqslant n/2$ (in fact, equality holds, by 1.15).

This example also illustrates the effect of allowing representations of any length when defining ν_1. If we were to allow only representations of length n -1 (= rank T), we would in fact only be considering representations of T_1.

If, in the above situation, there is a projection P of Y onto Y_1 and if $\Sigma f_i \otimes y_i$ is a representation of T, then $\Sigma f_i \otimes (Py_i)$ is a representation of T_1. Hence we have $\nu_1(T_1) \leqslant \|P\| \nu_1(T)$.

A further statement in the positive direction is:

1.18. Let T be in FL(X,Y) and $\varepsilon > 0$. Then there is a finite-dimensional subspace Y_1 of Y such that (with T_1 defined as above) $\nu_1(T_1) \leqslant (1+\varepsilon)\nu_1(T)$.

Proof. There is a representation $\Sigma f_i \otimes y_i$ for T with $\Sigma \|f_i\|.\|y_i\| \leqslant (1+\varepsilon)\nu_1(T)$. The statement follows on putting Y_1 = lin $(y_1, \dots, y_k)$.

Exercise. Let X be infinite-dimensional, and let S be an operator in FL(X*,Y) that can be represented in the form $\Sigma \hat{x}_i \otimes y_i$, where $x_i \in X$ (in other words, a weak-star continuous operator). Show that $\nu_1(S)$ can be obtained using only such representations. (By 1.18, we may assume Y finite-dimensional, hence Y = Z* and S = T* for some T in L(Z,X)).

Relationship with tensor products

We finish this section by describing (rather briefly) how the above results translate into statements about tensor products. This is not very important for our development, but it is a notion that is widely used in the literature.

The algebraic tensor product $X \otimes Y$ is the set of elements of the form $\sum_1^n x_i \otimes y_i$, which may be regarded as operators from X* to Y (or equally from Y* to X). If X or Y is finite-dimensional, it equates with the space of all such operators (and in any case, FL(X,Y) equates with $X^* \otimes Y$). Given a norm α on $X \otimes Y$, the tensor product $X \otimes_\alpha Y$ is defined to be the completion of $(X \otimes Y, \alpha)$.

If α is ordinary operator norm, the result of this is called the "injective" tensor product, usually denoted by $X \otimes_\varepsilon Y$ or $X \check{\otimes} Y$. The "projective" tensor product, denoted by $X \otimes_\gamma Y$ or $X \hat{\otimes} Y$, is derived from the norm $\gamma(u)$ = $\inf \{ \Sigma \|x_i\|.\|y_i\| : u = \Sigma x_i \otimes y_i \}$. For finite-dimensional X, this coincides with

ν_1, so our results on ν_1 can be regarded as results on this kind of tensor product. In particular, 1.11 equates the dual of $X \otimes_\gamma Y$ (for finite-dimensional X) with $X^* \otimes_\varepsilon Y^*$. Actually, there is a simple description of this dual that does not require finite-dimensionality, which we now describe (it will be used once, in section 10).

Let B(X,Y) be the space of all bilinear forms on X×Y, with norm defined by

$$\|\beta\| = \sup \{ |\beta(x,y)| : \|x\| \leq 1 , \|y\| \leq 1 \} .$$

There is a natural isometry between B(X,Y), L(X,Y*) and L(Y,X*), given by:

$$\beta(x,y) = (T_\beta x)(y) = (U_\beta y)(x) .$$

Exactly as in 1.3, one shows that if β is in B(X,Y) and u is in $X \otimes Y$, then the expresssion $\Sigma\, \beta(x_i,y_i)$ is independent of the representation $\Sigma\, x_i \otimes y_i$ for u and hence can be used to define a corresponding functional $\hat{\beta}$ on $X \otimes Y$. By essentially the same proof as 1.11, we now have:

<u>1.19 Proposition.</u> The dual of $(X \otimes_\gamma Y)$ is isometric to B(X,Y).

Proof. With the above notation, we have

$$|\hat{\beta}(u)| \leq \|\beta\| \,\Sigma\, \|x_i\| . \|y_i\| ,$$

and hence $\|\hat{\beta}\| \leq \|\beta\|$. For any $\varepsilon > 0$ there exist x_0,y_0 with $\|x_0\| = \|y_0\|$ and $|\beta(x_0,y_0)| = |\hat{\beta}(x_0 \otimes y_0)| \geq (1 - \varepsilon) \|\beta\|$. Hence $\|\hat{\beta}\| = \|\beta\|$. Finally, given any functional ϕ on $X \otimes_\gamma Y$, we have $\phi = \hat{\beta}$, where β is defined by $\beta(x,y) = \phi(x \otimes y)$.

It is instructive to see how this result translates back into the notation of 1.11. Since $(FL(X,Y),\nu_1)$ equates to $(X^* \otimes Y, \gamma)$, its dual can be identified with B(X*,Y), and hence with L(X*,Y*) or L(Y,X**). It is now clear why we needed reflexivity to equate this with L(Y,X).

While it might be said that this particular result has gained in both clarity and generality by the tensor product presentation, this is not an option when we come to the summing norms. These are defined for operators in terms of their action, and are in no way limited to finite-rank operators, or to operators approximable by finite-rank ones.

2. FINITE SEQUENCES OF ELEMENTS : THE QUANTITIES μ_1, μ_2

The definition and equivalent forms

The following notion is used in defining the summing and nuclear norms (other than ν_1). The results in this section are not deep, but they will greatly facilitate our later deliberations on these norms.

Let $p \geq 1$, and let $(x_1, \ldots, x_k)$ be a finite sequence of elements of a normed linear space X (real or complex). We define

$$\mu_p(x_1, \ldots, x_k) = \sup \{ (\Sigma\, |f(x_i)|^p)^{1/p} : f \in U_{X^*} \} .$$

(There is no generally accepted notation for this. Many writers do without a special notation, repeatedly writing out the right-hand side. Pietsch [OI] uses w_p where we use μ_p.)

Clearly, the ordering of the finite sequence makes no difference, and $\mu_p(x) = \|x\|$ for a "sequence" consisting of one element x. By the Hahn-Banach theorem, if the x_i belong to a subspace E of X, then $\mu_p(x_1, \ldots, x_k)$ is the same when evaluated in E and in X. Further immediate properties are summarized in the next result (the proofs are obvious).

<u>2.1</u> (i) $\mu_p(\alpha_1 x, \ldots, \alpha_k x) = (\Sigma\, |\alpha_i|^p)^{1/p}\, \|x\|$.

(ii) $\max \|x_i\| \leq \mu_p(x_1, \ldots, x_k) \leq (\Sigma\, \|x_i\|^p)^{1/p}$.

(iii) If $p \leq q$, then $\mu_p(S) \geq \mu_q(S)$ for any finite sequence S.

(iv) If T is an operator on X, then

$$\mu_p(Tx_1, \ldots, Tx_k) \leq \|T\|\, \mu_p(x_1, \ldots, x_k)$$

(v) $\mu_p(x_1, \ldots, x_n)^p \leq \mu_p(x_1, \ldots, x_k)^p + \mu_p(x_{k+1}, \ldots, x_n)^p$.

Note that "μ_∞" would be simply $\max \|x_i\|$. In fact, our real interest is in μ_1 and μ_2, and we proceed to look at these separately. We show that (1) the functionals in the definition can be restricted to a norming subset, and (2) the definitions can be formulated purely in terms of norms of linear combinations $\Sigma \lambda_i x_i$, avoiding the dual space.

2.2. Let $x_1, \dots ,x_k$ be elements of X, and let K be a norming subset of U_{X^*}. Let

$$A' = \sup \{ \Sigma |f(x_i| : f \in K \},$$

$$B = \sup \{ \|\Sigma \lambda_i x_i\| : |\lambda_i| \leqslant 1 \text{ for each } i\},$$

$$B' = \sup \{ \|\Sigma \alpha_i x_i\| : |\alpha_i| = 1 \text{ for each } i\}.$$

Then $A' = B = B' = \mu_1(x_1, \dots ,x_k)$.

Proof. Let $A = \mu_1(x_1, \dots ,x_k)$. Clearly, $A' \leqslant A$ and $B' \leqslant B$. We complete the proof by showing $B \leqslant A'$ and $A \leqslant B'$.

Let $|\lambda_i| \leqslant 1$ for each i, and take $\varepsilon > 0$. There exists $f \in K$ such that

$$(1-\varepsilon)\,\|\Sigma \lambda_i x_i\| \leqslant |f(\Sigma \lambda_i x_i)| = |\Sigma \lambda_i f(x_i)| \leqslant \Sigma |f(x_i)| \leqslant A'.$$

This shows that $(1-\varepsilon)B \leqslant A'$ for all $\varepsilon > 0$, so $B \leqslant A'$.

Now choose $f \in U_{X^*}$. For each i, take α_i with $|\alpha_i| = 1$ and $\alpha_i f(x_i) = |f(x_i)|$. Then

$$\Sigma\,|f(x_i)| = f(\Sigma \alpha_i x_i) \leqslant \|\Sigma \alpha_i x_i\| \leqslant B'.$$

Hence $A \leqslant B'$.

Remarks. (1) Hence in the real case, $\mu_1(x_1, \dots ,x_k)$ is the greatest of the 2^k numbers $\|\varepsilon_1 x_1 + \dots + \varepsilon_k x_k\|$, with each ε_i chosen from $\{-1, 1\}$. In the complex case, the supremum in B' is attained, by compactness, but not as the maximum of a finite set.

(2) If the supremum in B' is attained at $\Sigma\alpha_i x_i$, then the supremum in the original definition is attained by f satisfying $f(\Sigma\alpha_i x_i) = \|\Sigma\alpha_i x_i\|$ hence $\alpha_i f(x_i) = |f(x_i)|$.

2.3 Let $x_1, \dots ,x_k$ be elements of X, and let K be a norming subset of X^*. Let

$$C' = \sup \{ (\Sigma |f(x_i)|^2)^{1/2} : f \in K\},$$

$$D = \sup \{ \|\Sigma \lambda_i x_i\| : \Sigma|\lambda_i|^2 \leqslant 1\}.$$

Then $C' = D = \mu_2(x_1, \dots ,x_k)$.

Proof. Let $C = \mu_2(x_1, \ldots, x_k)$. Clearly $C' \leqslant C$. We show $C \leqslant D \leqslant C'$.

Let $\Sigma |\lambda_i|^2 \leqslant 1$ and $\varepsilon > 0$. There exists $f \in K$ such that

$$|\Sigma\ \lambda_i f(x_i)| \geqslant (1-\varepsilon)\ \|\Sigma \lambda_i x_i\| .$$

By Schwarz's inequality, $|\Sigma\ \lambda_i f(x_i)| \leqslant (\Sigma\,|f(x_i)|^2)^{1/2} \leqslant C'$. Hence $C' \geqslant (1-\varepsilon)D$ for all $\varepsilon > 0$, so $C' \geqslant D$.

Now choose $f \in U_{X^*}$. There exist numbers λ_i with $\Sigma\,|\lambda_i|^2 = 1$ and

$$(\Sigma |f(x_i)|^2)^{1/2} = \Sigma\ \lambda_i f(x_i) \leqslant \|\Sigma\ \lambda_i x_i\| \leqslant D .$$

Therefore $C \leqslant D$.

Hence we have

$$\|\ \Sigma \lambda_i x_i\| \leqslant (\Sigma\ |\lambda_i|^2)^{1/2}\ \mu_2(x_1, \ldots, x_k) .$$

In the same way, one finds that

$$\mu_p(x_1, \ldots, x_k) = \sup\{\ \|\Sigma\ \lambda_i x_i\| : \ \Sigma\ |\lambda_i|^{p'} \leqslant 1\} ,$$

where $\frac{1}{p} + \frac{1}{p'} = 1$.

<u>2.4.</u> For $f_1, \ldots, f_k$ in X^* and $p = 1,2,$

$$\mu_p(f_1, \ldots, f_k) = \sup\ \{\ (\Sigma\,|f_i(x)|^p)^{1/p}\ :\ x \in U_X\}$$

Proof. U_X (or strictly, the corresponding set in X^{**}) is a norming set of functionals on X^*.

We can now show that the quantities μ_p equate with the norms of certain operators having the elements as "rows" or "columns."

<u>2.5</u> (i) For S in $L(\ell_\infty^k, X)$, $\|S\| = \mu_1(Se_1, \ldots, Se_k)$.

(ii) For S in $L(\ell_2^k, X)$, $\|S\| = \mu_2(Se_1, \ldots, Se_k)$.

(iii) Let T in $L(X, \ell_p^k)$ be given by : $Tx = \sum_i f_i(x) e_i$, where $f_i \in X^*$. Then $\|T\| = \mu_p(f_1, \ldots, f_k)$ ($p = 1$ or 2).

Proof. (i) We have

$$\|S\| = \sup\ \{\ \|Sx\| : \ \|x\|_\infty \leqslant 1\}$$
$$= \sup\ \{\ \|\ \Sigma\ x(i)(Se_i)\| : \|x\|_\infty \leqslant 1\} .$$

This equals $\mu_1(Se_1, \ldots, Se_k)$, by 2.2.

(ii) Similar to (i), using 2.3.

(iii) This follows at once from 2.4.

Exercise. Show that

$$\mu_1(\alpha_1 x_1, \ldots, \alpha_k x_k) \leqslant (\Sigma |\alpha_i|^2)^{1/2} \mu_2(x_1, \ldots, x_k).$$

Some particular cases

For spaces of the form $\ell_\infty(S)$ (in particular, ℓ_∞^n) or $C(K)$, there is a very simple interpretation of μ_p:

2.6. In $\ell_\infty(S)$ or $C(K)$,

$$\mu_p(x_1, \ldots, x_k)^p = \|\Sigma |x_i|^p\| \qquad \text{for } p = 1,2.$$

Proof. The point-evaluation functionals $\delta_s(x) = x(s)$ form a norming set. The statement follows, by 2.2, 2.3.

In particular, if elements x_i of $\ell_\infty(S)$ have disjoint support, then

$$\mu_1(x_1, \ldots, x_k) = \mu_2(x_1, \ldots, x_k) = \max \|x_i\| .$$

2.7. If $x_1, \ldots, x_k$ are positive elements of ℓ_1^n (or ℓ_1), then

$$\mu_1(x_1, \ldots, x_k) = \Sigma \|x_i\| ,$$

$$\mu_2(x_1, \ldots, x_k) = (\Sigma \|x_i\|^2)^{1/2} .$$

Proof. Let f be the functional defined by $f(x) = \Sigma\, x(i)$. Then $\|f\| = 1$ and $f(x) = \|x\|$ for positive x. The statements follow.

2.8. Let $e_1, \ldots, e_k$ be orthonormal elements of an inner-product space. Then

$$\mu_1(\alpha_1 e_1, \ldots, \alpha_k e_k) = (\Sigma |\alpha_i|^2)^{1/2} ,$$

$$\mu_2(\alpha_1 e_1, \ldots, \alpha_k e_k) = \max |\alpha_i|.$$

Proof. Both statements follow at once from the equality

$$\|\Sigma \lambda_i \alpha_i e_i\|^2 = \Sigma\, |\lambda_i \alpha_i|^2 ,$$

together with 2.2 and 2.3.

The last three results enable us to tabulate the values of $\mu_p(e_1, \dots, e_n)$ in ℓ_∞^n, ℓ_2^n, ℓ_1^n :

	ℓ_∞^n	ℓ_2^n	ℓ_1^n
μ_1	1	$\sqrt{n}$	n
μ_2	1	1	$\sqrt{n}$.

2.9 Example. Let $x_1 = (1,1)$, $x_2 = (1,-1)$. In the space ℓ_1^2 , we have $\|x_1 + x_2\| = \|x_1 - x_2\| = 2$, so $\mu_1(x_1,x_2) = 2$. Since $\|x_1\| = 2$, we also have $\mu_2(x_1,x_2) = 2$.

The reader may care to calculate $\mu_p(x_1,x_2)$ in ℓ_2^2 and ℓ_∞^2 .

2.10. If $x_1, \dots, x_k$ are positive elements of any normed lattice, then $\mu_1(x_1, \dots, x_k) = \| \Sigma x_i \|$.

Proof. If $|\lambda_i| \leqslant 1$ for each i, then $|\Sigma \lambda_i x_i| \leqslant \Sigma x_i$, so we have $\|\Sigma \lambda_i x_i\| \leqslant \|\Sigma x_i\|$.

We now state one result on interchanging the rows and columns of a matrix. Given elements $a_1, \dots, a_m$ of $\mathbb{R}^n$, we can define elements $\tilde{a}_1, \dots, \tilde{a}_n$ of $\mathbb{R}^m$ by : $\tilde{a}_j(i) = a_i(j)$. We use the notation $\mu_p^{(q)}$ to mean μ_p with respect to $\| \ \|_q$.

2.11. With this notation, we have

$$\mu_1^{(1)} (\tilde{a}_1, \dots, \tilde{a}_n) = \mu_1^{(1)}(a_1, \dots, a_m) .$$

Proof. Let f be the functional on $\mathbb{R}^n$ defined by $f(a) = \sum_j t_j a(j)$. Then

$$\sum_i |f(a_i)| = \sum_i |\sum_j t_j a_i(j)| = \sum_i |\sum_j t_j \tilde{a}_j(i)| = \|\sum_i t_j \tilde{a}_j\|_1 .$$

The statement follows on considering the supremum for all choices of t_j with $|t_j| \leqslant 1$.

(Alternatively, one can deduce 2.11 from 2.5).

There are further results of this kind, though we will not require them. For example:

Exercise. Show that $\mu_2^{(1)}(\tilde{a}_1, \dots, \tilde{a}_n) = \mu_1^{(2)}(a_1, \dots, a_m)$.

2.12 Proposition. For elements x_i of an inner-product space,

$$\mu_1(x_1, \dots, x_k) \geq (\sum_i \|x_i\|^2)^{\frac{1}{2}}.$$

Proof. We show by induction that there exist $\varepsilon_1, \dots, \varepsilon_k$ in $\{-1,1\}$ such that $\|\Sigma\, \varepsilon_i x_i\|^2 \geq \Sigma\, \|x_i\|^2$. Assume this statement for sequences of length k, and let $x_1, \dots, x_{k+1}$ be given. Choose $\varepsilon_1, \dots, \varepsilon_k$ as just stated, and write $y = \sum_1^k \varepsilon_i x_i$. The statement for k+1 follows at once from the equality

$$\|y + x_{k+1}\|^2 + \|y - x_{k+1}\|^2 = 2\|y\|^2 + 2\|x_{k+1}\|^2 .$$

We will return to this property later. We now look briefly at the problem of identifying the scalars α_i such that $\|\Sigma\, \alpha_i x_i\|$ is the element at which μ_1 or μ_2 is attained. We include this because it is illuminating, but it will not be used again.

2.13. Let $x_1, \dots, x_k$ be elements of an inner-product space, and let $\alpha_1, \dots, \alpha_k$ be scalars such that $|\alpha_i| = 1$ and

$$\|\Sigma\, \alpha_i x_i\| = \mu_1(x_1, \dots, x_k) .$$

For $F \subseteq \{1, \dots, k\}$, write $S_F = \Sigma\, \{\alpha_i x_i : i \in F\}$, and let F' be the complement of F. Then $\langle S_F, S_{F'} \rangle$ is real and non-negative for each F. The converse is true in the real case.

Proof. Clearly, $\|S_F + S_{F'}\| \geq \|S_F + \alpha S_{F'}\|$ for all α with $|\alpha| = 1$ (and in the real case, this is sufficient). For real scalars, the result follows at once by taking $\alpha = -1$. In the complex case, choose α such that $\bar{\alpha} \langle S_F, S_{F'} \rangle = |\langle S_F, S_{F'} \rangle|$. Then

$$\|S_F + \alpha S_{F'}\|^2 = \|S_F\|^2 + \|S_{F'}\|^2 + 2|\langle S_F, S_{F'} \rangle| .$$

The statement follows.

2.14 Example. In real ℓ_2^2, let

$$x_1 = (1, 0), \quad x_2 = (\tfrac{1}{2}, \tfrac{\sqrt{3}}{2}), \quad x_3 = (\tfrac{1}{2}, -\tfrac{\sqrt{3}}{2}),$$

so that $\langle x_1, x_2 \rangle = \langle x_1, x_3 \rangle = \frac{1}{2}$ and $\langle x_2, x_3 \rangle = -\frac{1}{2}$. By 2.13,

$$\mu_1(x_1, x_2, x_3) = \|x_1 + x_2 + x_3\| = 2.$$

The choice of scalars λ_i such that $\| \Sigma \lambda_i x_i \| = \mu_2(x_1, \ldots, x_k)$ can be described as follows. Let A be the matrix $(\langle x_i, x_j \rangle)$. If $y = \Sigma\, t(i) x_i$, then $\langle y, y \rangle = \langle At, t \rangle$. Hence we require the t with $\|t\|_2 = 1$ for which $\langle At, t \rangle$ is largest. By the theory of positive definite operators, this occurs when t is an eigenvector corresponding to the largest eigenvalue λ_0 of A, and we have

$$\mu_2(x_1, \ldots, x_k)^2 = \lambda_0 = \|A\| .$$

A note on complex scalars

Any linear space X over $\mathbb{C}$ can, of course, be regarded as a linear space over $\mathbb{R}$ (which we denote by X_R). This leads to competing definitions of μ_p, which we denote (temporarily) by μ_p^C and μ_p^R. When formulated in terms of $\|\Sigma \lambda_j x_j\|$, μ_p^C allows complex scalars, while μ_p^R only allows real ones. So clearly $\mu_p^C(S) \geqslant \mu_p^R(S)$ for any S. Conversely, it is easy to see that

$$\mu^C(S) \leqslant 2\mu^R(S), \qquad \mu^C(S) \leqslant \sqrt{2}\, \mu^R(S)$$

(the second statement is most easily seen from the dual formulation of μ_2). To give examples of inequality, one need go no further than $\mathbb{C}$ itself; the corresponding real space is ℓ_2^2, and one has, for example,

$$\mu_1^C(1, i) = 2, \qquad \mu_1^R(1, i) = \sqrt{2} .$$

The comparison has more point when elements of real ℓ_p^n, C(S), etc. are regarded as elements of the corresponding complex space. In $\ell_\infty(S)$ or C(S), 2.6 shows that μ_p^C and μ_p^R then coincide. We indicate briefly by examples what can happen in other cases.

<u>2.15 Example.</u> Let x_1, x_2 be as in 2.9. We saw that $\mu_1^R(x_1,x_2) = 2$. However, $\|x_1 + ix_2\| = 2\sqrt{2}$, so $\mu_1^C(x_1,x_2) \geqslant 2\sqrt{2}$.

<u>2.16 Example.</u> Let H be a complex Hilbert space, and let $x_1, \ldots, x_k$ be elements such that $\langle x_i, x_j \rangle$ is real for each i,j. Then $\mu_2^C(x_1, \ldots, x_k) = \mu_2^R(x_1, \ldots, x_k)$. (This follows easily from the fact that $\|u+iv\|^2 = \|u\|^2 + \|v\|^2$ when $\langle u,v \rangle$ is real).

<u>Exercise.</u> Consider 2.14 again. Use 2.13 to show that $\mu_1^C(x_1,x_2,x_3)$ is attained at $x_1 + \alpha x_2 + \bar{\alpha} x_3$, where $\alpha = \frac{1}{2} + \frac{\sqrt{3}}{2} i$, having the value $3/\sqrt{2}$. (This shows that the converse implication in 2.13 is not true in the complex case).

3. THE SUMMING NORMS

The definition and immediate consequences

Let T be an operator between normed linear spaces. For (finite) $p \geq 1$, the p-summing norm π_p is defined by

$$\pi_p(T) = \sup \left\{ \left(\sum_i \|Tx_i\|^p \right)^{1/p} : \mu_p(x_1, \ldots, x_k) \leq 1 \right\},$$

where μ_p is defined as in section 2. In this, all finite sequences (x_i) are considered: there is no restriction on the length k. Clearly, for any elements x_i, we have (for example) :

$$\sum \|Tx_i\| \leq \pi_1(T)\, \mu_1(x_1, \ldots, x_k) .$$

The operator T is said to be "p-summing" if $\pi_p(T)$ is finite (some writers insert - somewhat unnecessarily - the word "absolutely"). We denote $P_p(X,Y)$ the set of all p-summing operators from X to Y. Note that "π_∞" would be ordinary operator norm.

Generally, we shall present our results in terms of π_1 and π_2, rather than general p. These are the "natural" cases, and, as we shall see, each has some special characteristics of its own. (Where appropriate, the reader may treat it as an exercise to fill in the details for other p).

3.1. $P_p(X,Y)$ is a linear subspace of L(X,Y), and π_p is a norm on it. Further:

$$\pi_p(T) \geq \|T\| ,$$
$$\pi_p(BT) \leq \|B\|\, \pi_p(T) ,$$
$$\pi_p(TA) \leq \pi_p(T)\, \|A\| .$$

Proof. Clearly, $\pi_p(T) \geq \|T\|$, since $\mu_p(x) = \|x\|$.

The only property whose verification is non-trivial is $\pi_p(S + T) \leq \pi_p(S) + \pi_p(T)$. We give the proof of this for the case $p = 2$.

Let $\mu_2(x_1, \dots, x_k) \leq 1$. Write $\|Sx_i\| = \alpha_i$, $\|Tx_i\| = \beta_i$, $\|(S+T)x_i\| = \gamma_i$. Then $\gamma_i \leq \alpha_i + \beta_i$, so

$$\begin{aligned}(\Sigma\gamma_i^2)^{1/2} &\leq [\Sigma(\alpha_i + \beta_i)^2]^{1/2} \\ &\leq (\Sigma\alpha_i^2)^{1/2} + (\Sigma\beta_i^2)^{1/2} \qquad \text{(by Minkowski's inequality)} \\ &\leq \pi_2(S) + \pi_2(T).\end{aligned}$$

The following very simple result is of great help in understanding the meaning of the summing norms, and in obtaining an estimate for them in particular cases. As we shall see in section 5, it can actually serve as an alternative definition in the finite dimensional case.

3.2 <u>Proposition</u>. Let $p \geq 1$. Suppose that there are functionals $f_1, f_2, \dots$ (a finite or infinite sequence) such that $\|Tx\|^p \leq \sum_j |f_j(x)|^p$ for all x. Then

$$\pi_p(T) \leq (\Sigma \|f_j\|^p)^{1/p}.$$

Proof. Let $\mu_p(x_1, \dots, x_k) \leq 1$. Then

$$\sum_i |f_j(x_i)|^p \leq \|f_j\|^p \qquad \text{for each } j,$$

so

$$\sum_i \|Tx_i\|^p \leq \sum_i \sum_j |f_j(x_i)|^p \leq \sum_j \|f_j\|^p,$$

In particular, $\pi_1(T) \leq \nu_1(T)$ for finite-rank operators.

<u>3.3.</u> If T is 1-summing, then it is 2-summing, and $\pi_2(T) \leq \pi_1(T)$.

Proof. Let $\mu_2(x_1, \dots, x_k) \leq 1$, and choose positive scalars λ_i. For f in U_{X^*}, we have

$$\sum_i \lambda_i |f(x_i)| \leq (\sum_i \lambda_i^2)^{1/2},$$

since $\sum_i |f(x_i)|^2 \leq 1$. Hence $\mu_1(\lambda_1 x_1, \dots, \lambda_k x_k) \leq (\sum_i \lambda_i^2)^{1/2}$, so

$$\sum_i \lambda_i \|Tx_i\| \leq (\sum_i \lambda_i^2)^{1/2} \pi_1(T).$$

The result follows on putting $\lambda_i = \|Tx_i\|$.

Similarly, of course, one can show that $\pi_q(T) \leqslant \pi_p(T)$ for $p \leqslant q$.

We now list a number of immediate consequences of the definition and the above results.

(1) Since π_p is defined in terms of the action of T on finite sets of elements, we have:

$$\pi_p(T) = \sup \{ \pi_p(T|_E) : \text{ E a finite-dimensional subspace of X } \} .$$

Also, if (E_n) is a sequence of subspaces with $\bigcup_{n=1}^{\infty} E_n$ dense in X, then $\pi_p(T) = \sup \pi_p(T|_{E_n})$.

(2) If T is in L(X,Y) and Y is a subspace of Z, then $\pi_p(T)$ (unlike nuclear norm) is the same when T is regarded as an element of L(X,Z).

(3) For a rank one operator $T = f \otimes y$, we have (using 3.2)

$$\pi_p(T) = \|T\| = \|f\| . \|y\| \qquad \text{for any p.}$$

Hence π_p is an operator ideal norm.

(4) If dim X = n, then, by 3.2 and 3.3,

$$\pi_2(I_X) \leqslant \pi_1(I_X) \leqslant \nu_1(I_X) = n .$$

Hence all operators of rank n are 1-summing, with $\pi_1(T) \leqslant n\|T\|$ (since $T = I_Y T$).

For a finite-dimensional space X, $\pi_1(I_X)$ can be regarded as a constant in some way descriptive of the space; sometimes we write $\pi_1(X)$ instead of $\pi_1(I_X)$. Clearly for isomorphic spaces X, Y, we have

$$\pi_1(Y) \leqslant d(X,Y)\, \pi_1(X) .$$

In principle, the same is true of $\pi_2(I_X)$, but we shall see that in fact this equals $\sqrt{n}$ for every n-dimensional space (and hence that I_X is never 2-summing when X is infinite-dimensional).

If T is an isometry of X onto itself, then it is clear from the identities $T = TI_X$ and $I_X = T^{-1}T$ that $\pi_p(T) = \pi_p(I_X)$.

(5) In the definition of π_p, finite sequences of any length are allowed. If the definition is modified by restricting the length to a certain n, then we obtain a norm which we denote by $\pi_p^{(n)}$. Clearly $\pi_p^{(n)}(T)$ increases with n and tends to $\pi_p(T)$ as $n \to \infty$. For operators of rank n, one might expect $\pi_p^{(n)}(T)$ to bear some relation to $\pi_p(T)$. Later examples and results will reveal the extent to which this is the case.

(6) It is routine to verify that if Y is complete, then so is $[P_p(X,Y), \pi_p]$ (though this is not important for our purposes).

In general it is by no means easy to calculate $\pi_p(T)$ for a particular operator T, even when T is quite simple. In this section, we will describe a few cases in which it does happen to be easy. Some of the results obtained are not just included for practice: they will be used in the ensuing general theory.

For now, consider diagonal operators in $\mathbb{R}^n$. As before, $I^{(n)}$ denotes the identity in $\mathbb{R}^n$, and if T is an operator from $\mathbb{R}^n$ to R^m, then $T_{p,q}$ denotes T regarded as a mapping from ℓ_p^n to ℓ_q^m.

3.5. Let T be the operator $\operatorname{diag}(\alpha_1, \ldots, \alpha_n)$ in $\mathbb{R}^n$. If each of p, q is either 2 or ∞, then $\pi_2(T_{p,q}) = (\Sigma\, \alpha_i^2)^{1/2}$. In particular, $\pi_2(I_{p,q}^{(n)}) = \sqrt{n}$. (Similarly for $\mathbb{C}^n$).

Proof. In ℓ_2^n and ℓ_∞^n, we have $\mu_2(e_1, \ldots, e_n) = 1$. Hence

$$\pi_2(T_{p,q})^2 \geq \Sigma\, \|Te_i\|^2 = \Sigma\, \alpha_i^2 .$$

For x in $\mathbb{R}^n$,

$$\|Tx\|_\infty^2 \leq \|Tx\|_2^2 = \Sigma\, \alpha_i^2 x(i)^2 .$$

This is of the form $\Sigma\, f_i(x)^2$, where $\|f_i\| = \alpha_i$. By 3.2, it follows that $\pi_2(T_{p,q})^2 \leq \Sigma\, \alpha_i^2$.

Note. We have shown that in this case, $\pi_2(T) = \pi_2^{(n)}(T)$.

The premultiplication lemma and $\pi_2(I_X)$

3.6. (i) Let T be in $P_1(X,Y)$ and $\varepsilon > 0$. Then there exist k and an operator A in $L(\ell_\infty^k, X)$ such that $\|A\| = 1$ and $\pi_1(TA) \geq (1-\varepsilon)\, \pi_1(T)$.

(ii) Let T be in $P_2(X,Y)$ and $\varepsilon > 0$. Then there exist k and an operator A in $L(\ell_2^k, X)$ such that $\|A\| = 1$ and $\pi_2(TA) \geq (1-\varepsilon)\, \pi_2(T)$. Further, if rank T = n, we may take $k \leq n$.

Proof. (i) There exist elements x_i of X with $\mu_1(x_1, \ldots, x_k) = 1$ and $\Sigma\, \|Tx_i\| \geq (1-\varepsilon)\, \pi_1(T)$. Define $A : \ell_\infty^k \to X$ by: $Ae_i = x_i$. Then $\|A\| = 1$ by 2.5) and $\Sigma\, \|TAe_i\| \geq (1-\varepsilon)\, \pi_1(T)$. Since $\mu_1(e_1, \ldots, e_k) = 1$ in

ℓ_∞^k, we have $\pi_1(TA) \geqslant (1-\varepsilon)\,\pi_1(T)$.

(ii) The first statement is proved as in (i). Suppose that rank $T = n$. Let $N = \ker(TA)$, $H = N^{\perp}$ and let P be the orthogonal projection onto H. Write $A_1 = A|_H$. Then TA_1 is one-to-one, so $\dim H \leqslant n$. Also, $TA(x - Px) = 0$ for all x in ℓ_2^k, so $TA = TA_1P$. Hence $\pi_2(TA) \leqslant \pi_2(TA_1)$, and A_1 is the required mapping.

3.7 Corollary. If rank $T = n$, then

$$\pi_2(T) = \sup\,\{\pi_2(T|_E) : \dim E = n\}.$$

Proof. Take $E = A_1(H)$.

3.8 Proposition. Let X be any n-dimensional normed linear space. Then $\pi_2(I_X) \leqslant \sqrt{n}$.

Proof. By 3.6, there exist a Hilbert space H with $\dim H \leqslant n$ and an operator $A : H \to X$ with $\|A\| = 1$ and $\pi_2(A) \geqslant (1-\varepsilon)\,\pi_2(I_X)$. By 3.5, $\pi_2(I_H) \leqslant \sqrt{n}$. Since $A = AI_H$, we have

$$\pi_2(A) \leqslant \|A\|\,\sqrt{n} = \sqrt{n}.$$

It follows that if rank $T = n$, then $\pi_2(T) \leqslant \sqrt{n}\,\|T\|$.

We shall see in section 5 that in fact $\pi_2(I_X) = \sqrt{n}$ for any n-dimensional space X.

Operators between Hilbert spaces

For any operator T between Hilbert spaces, there is a very satisfactory characterisation of $\pi_2(T)$:

3.9 Proposition. Let H_1, H_2 be Hilbert spaces (of finite or infinite dimension), and let T be a 2-summing operator from H_1 to H_2. Then:

(i) $\pi_2(T)^2 = \sum_i \|Te_i\|^2$ for any orthonormal base (e_i) of H_1;

(ii) $\pi_2(T^*) = \pi_2(T)$.

Proof. For any finite set F of indices, we have $\mu_2\{e_i : i \in F\} = 1$, and hence $\sum_{i\in F} \|Te_i\|^2 \leqslant \pi_2(T)^2$. Therefore $\sum_i \|Te_i\|^2 \leqslant \pi_2(T)^2$ (in the sense of summation if (e_i) is uncountable).

For y in H_2, we have $\|T^*y\|^2 = \sum_i \langle T^*y, e_i \rangle^2 = \sum_i \langle y, Te_i \rangle^2$.

Hence by 3.2,

$$\pi_2(T^*)^2 \leq \sum_i \|Te_i\|^2 \leq \pi_2(T)^2 .$$

Now let (f_j) be an orthonormal base of H_2. For x in H_1, we have $\|Tx\|^2 \leq \sum_j \langle Tx, f_j \rangle^2 = \sum_j \langle x, T^*f_j \rangle^2$. Hence, in the same way,

$$\pi_2(T)^2 \leq \sum_j \|T^*f_j\| \leq \pi_2(T^*)^2 .$$

The statements follow. (Of course, the complex case requires the insertion of modulus signs.)

In other words, for operators between Hilbert spaces, π_2 coincides with the "Hilbert-Schmidt" norm.

Clearly, we have shown in 3.9 that if $\dim H_1 = n$, then $\pi_2(T) = \pi_2^{(n)}(T)$.

It follows from 3.9 that an inner product (inducing the norm π_2) can be defined on $P_2(H_1, H_2)$ by setting

$$\langle S, T \rangle = \sum_i \langle Se_i. Te_i \rangle \quad (= \text{trace } T^*S) ,$$

this sum being independent of the choice of orthonormal base.

Note that for the operator $T : \ell_2^n \to \ell_2^m$ with matrix $(\alpha_{i,j})$, we have

$$\pi_2(T)^2 = \sum_i \sum_j |\alpha_{i,j}|^2 .$$

In the proof of 3.9, we have in fact established the following statement for the cases where only one of the two spaces is a Hilbert space :

3.10. Let H_1, H_2 be Hilbert spaces, with orthonormal bases (e_i), (f_j) respectively. Let X, Y be normed linear spaces. Then :

(i) for S in $P_2(H_1, Y)$,

$$\pi_2(S^*)^2 \leq \sum_i \|Se_i\|^2 \leq \pi_2(S)^2 ;$$

(ii) for T in $P_2(X, H_2)$,

$$\pi_2(T)^2 \leq \sum_j \|T^*f_j\|^2 \leq \pi_2(T^*)^2 .$$

3.11 Example. Let $S = I_{2,\infty}^{(2)}$, $T = I_{1,2}^{(2)}$. Then $S = T^*$. We have seen (3.5) that $\pi_2(S) = \sqrt{2}$. We show that $\pi_2(T) = 1$, establishing that inequality can hold in both statements in 3.10. (We shall see in section 6 that in fact $\pi_2(I_{1,2}^{(n)}) = 1$ for all n.)

Consider the orthonormal base (f_1, f_2), where $f_1 = \frac{1}{\sqrt{2}}(1, 1)$, $f_2 = \frac{1}{\sqrt{2}}(1, -1)$. By 3.10, $\pi_2(T)^2 \leq \|Sf_1\|^2 + \|Sf_2\|^2 = 1$. Equality holds, since $\|T\| = 1$.

This example also shows that in the situation of 3.10, $\Sigma \|Se_i\|^2$ and $\Sigma \|T^*f_j\|^2$ are not independent of the choice of base.

We mention the following easy consequence of 3.9, though it is not central to our development:

3.12. Let H_1, H_2 be infinite-dimensional Hilbert spaces. Then the finite-rank operators are dense in $[P_2(H_1, H_2), \pi_2]$. Hence every 2-summing operator between Hilbert spaces is compact.

Proof. Let (e_α) be an orthonormal base of H_1. By 3.9, the set of e_α with $Te_\alpha \neq 0$ is countable, so can be expressed as a sequence (e_n). Let

$$T_n x = \sum_{i=1}^{n} \langle x, e_i \rangle Te_i .$$

Then

$$\pi_2(T - T_n)^2 = \sum_{n=1}^{\infty} \|Te_i\|^2 \to 0 \qquad \text{as } n \to \infty .$$

3.13 Example. Define T in $L(\ell_2)$ by : $(Tx)(n) = \frac{1}{\sqrt{n}} x(n)$ for all n. Then T is compact, but not 2-summing, since $\Sigma \|Te_i\|^2$ is divergent.

Operators on L_∞-spaces

For operators on ℓ_∞^n, there is a very simple - and useful - characterisation of π_1 :

3.14. Let T be an operator from ℓ_∞^n to any normed linear space Y. Then

$$\pi_1(T) = \pi_1^{(n)}(T) = \nu_1(T) = \sum_i \|Te_i\| .$$

Proof. Since $\mu_1(e_1, \ldots, e_n) = 1$, we have $\pi_1^{(n)}(T) \geq \Sigma \|Te_i\|$. Conversely, the natural expression $T = \Sigma\, e_i \otimes (Te_i)$ gives $\nu_1(T) \leq \Sigma \|Te_i\|$.

This re-proves 1.15. Also, it shows that if $T = \text{diag}(\alpha_1, \ldots, \alpha_n)$, then $\pi_1(T_{\infty,q}) = \Sigma |\alpha_i|$ for each $q \geq 1$.

Exercise. For an operator T defined on c_0, prove that $\pi_1(T) = \sum_n \|Te_n\|$.

We can deduce a general relationship between π_1, ν_1 and projection constants.

3.15. If X is finite-dimensional and T is in L(X,Y), then

$$\nu_1(T) \leq \lambda(X)\, \pi_1(T).$$

Proof. It is enough to prove this for the case where X is a subspace of ℓ_∞^k for some k, (by 0.13, X is nearly isometric to such a subspace). Let P be a projection of ℓ_∞^k onto X. Then TP maps from ℓ_∞^k to Y and extends T. Hence, by 3.14,

$$\nu_1(T) \leq \nu_1(TP) = \pi_1(TP) \leq \pi_1(T)\, \|P\|.$$

3.16 Corollary. If dim X = n, then $\lambda(X)\, \pi_1(X) \geq n$.

Proof. Apply 3.15 with $T = I_X$.

In the following, L_∞ denotes any of the spaces ℓ_∞^n, $\ell_\infty(S)$, C(K), $L_\infty(\mu)$. There is a useful variant of 3.2 for operators defined on a subspace of L_∞. This is really quite general, since every normed linear space can be regarded as a subspace of $\ell_\infty(S)$ for a suitable S (see 0.12). We use the fact that L_∞ is both a lattice and an algebra : given an element x, there are well-defined elements $|x|$ and x^2.

3.17 Let X be a subspace of L_∞, and let T be an operator defined on X. Suppose that there is a positive linear functional ϕ on L_∞ such that $\|Tx\|^p \leq \phi(|x|^p)$ for all x in X. Then $\pi_p(T)^p \leq \|\phi\|$.

Proof. Choose elements x_i of X. Recall from 2.6 that

$$\mu_p(x_1, \dots, x_k)^p = \| \Sigma |x|^p \|$$

Assume that the value of this is not greater than 1. Then

$$\sum_i \|Tx_i\|^p \leq \sum_i \phi(|x_i|^p) = \phi(\sum_i |x_i|^p) \leq \|\phi\| .$$

Notes (1) There is no requirement here that $|x|^p$ belongs to X when x does. But clearly, if this condition is satisfied, then it is enough for ϕ to be defined on X.

(2) In the complex case, it is sufficient if ϕ is defined on real L_∞. However, if it is defined on complex L_∞, the meaning of "positive" is that $\phi(x)$ is real and positive whenever x is real and positive.

The converse of 3.17 is in fact true as well. This is Pietsch's theorem, the fundamental theorem on p-summing operators. We give an account of it in section 5.

From 3.17, we obtain at once one of the classical infinite-dimensional examples of p-summing operators:

3.18 Example. Let $I = [0,1]$, and let J_p be the formal identity mapping from $C(I)$ to $L_p(I)$. Then $\pi_p(J_p) = 1$ for $p = 1, 2$.

Proof. Let $\phi(x) = \int_0^1 x$. Then $\|\phi\| = 1$ (as a functional on $C(I)$) and $\|J_p x\|^p = \phi(|x|^p)$.

In the same way, if g is in $L_1(S,\mu)$ and T_g is the multipication operator from $C(S)$ to $L_1(S,\mu)$ given by $T_g(f) = fg$, then $\pi_1(T_g) = \|T_g\| = \|g\|_1$.

It is not very hard to show that J_2 is not 1-summing. However, an easier example is as follows :

3.19 Example. Define $T : \ell_\infty \to \ell_2$ by $(Tx)(n) = \frac{1}{n} x(n)$. Then $\|Tx\|^2 = \Sigma \frac{1}{n^2} x(n)^2$. This equals $\phi(x^2)$, where ϕ is a positive functional having $\|\phi\| = \Sigma \frac{1}{n^2}$. Hence T is 2-summing. It is not 1-summing, since $\mu_1(e_1, \dots, e_n) = 1$ for all n, while $\Sigma \|Te_i\|$ is divergent.

We now establish some nice facts about positive operators defined on (real) L_∞ or a sublattice (that is, a subspace X such that $|x|$ belongs to X whenever x does). For a positive operator on any normed lattice, we have that if $|x| \leq y$, then $|Tx| \leq Ty$ (since $-Ty \leq Tx \leq Ty$), hence $\|Tx\| \leq \|Ty\|$. In particular, $\|Tx\| \leq \|T(|x|)\|$.

Let e be the element of L_∞ with constant value 1. The unit ball of L_∞ is the set of x satisfying $|x| \leq e$. Hence for a positive operator (or functional) on L_∞, we have $\|T\| = \|Te\|$.

3.20. Let L_1 denote any of the spaces ℓ_1^n, ℓ_1, $L_1(\mu)$. Let X be a linear sublattice of L_∞, and let T be a positive operator from X into L_1. Then $\pi_1(T) = \|T\|$.

Proof. There is a linear functional ψ on L_1 such that $\|\psi\| = 1$ and $\psi(y) = \|y\|$ for all $y \geq 0$ (this is the characteristic property of L_1 as a normed lattice). For $x \in X$, we have

$$\|Tx\| \leq \|T(|x|)\| = \psi[T(|x|)] = (T^*\psi)(|x|).$$

But $\|T^*\psi\| \leq \|T\|$, so the statement follows, by 3.17.

3.21 Example. Let T be the operator from ℓ_∞^2 to ℓ_1^2 defined by : $Te_1 = (1,1)$, $Te_2 = (1,-1)$. Then $\|T\| = 2$ (e.g. by 2.9), while $\pi_1(T) = \|Te_1\| + \|Te_2\| = 4$. So 3.20 certainly does not hold for non-positive operators.

We now show, by a different method, that a similar statement holds when ℓ_1, π_1 are replaced by ℓ_2, π_2.

3.22. Let X be a linear sublattice of L_∞, and let T be a positive operator from X into ℓ_2^n or ℓ_2. Then $\pi_2(T) = \|T\|$.

Proof. Let $Tx = \sum_j f_j(x)e_j$, so that $\|Tx\|^2 = \sum_j f_j(x)^2$. (Remark: $f_j = T^*e_j$). By 3.2,

$$\pi_2(T)^2 \leq \sum_j \|f_j\|^2 .$$

If X contains e, the statement follows at once, since

$$\sum_j \|f_j\|^2 = \sum_j f_j(e)^2 = \|Te\|^2 = \|T\|^2 .$$

We now show that this equality still holds when X does not contain e. We must show that $\sum_1^N \|f_j\|^2 \leqslant \|T\|^2$ for each N. Take $\varepsilon > 0$. For each j, there is a positive element x_j of the unit ball such that $f_j(x_j) \geqslant (1-\varepsilon)\ \|f_j\|$. Let $y = \sup(x_1, \ldots, x_N)$. Then $\|y\| \leqslant 1$ and $f_j(y) \geqslant f_j(x_j)$ for each j, so

$$\|Ty\|^2 = \sum_1^N f_j(y)^2 \geqslant (1-\varepsilon)^2 \sum_1^N \|f_j\|^2 .$$

Exercise. Let T be a positive operator from (the whole of) L_∞ to ℓ_2. Use the fact that $(x - ce)^2 \geqslant 0$ to show that $f(x)^2 \leqslant f(x^2)f(e)$ for a positive functional f. Deduce the existence of a functional ϕ as in 3.17.

Finally, we show that for operators on L_∞ (or a sublattice), the norms π_p are "monotonic". Simple examples show that this is not true for operators on other normed lattices.

Define (temporarily) $\pi_p^+(T)$ in the same way as $\pi_p(T)$, except that only positive elements x_i are allowed. (We come back to this idea in section 18). Clearly, if $0 \leqslant S \leqslant T$, then $\pi_p^+(S) \leqslant \pi_p^+(T)$.

3.23. Let X be a sublattice of L_∞, Y any normed lattice. For positive operators from X to Y, $\pi_p(T) = \pi_p^+(T)$. Hence if $0 \leqslant S \leqslant T$, then $\pi_p(S) \leqslant \pi_p(T)$.

Proof. This follows at once from the fact (special to L_∞!) that $\mu_p(|x_1|, \ldots, |x_k|) = \mu_p(x_1, \ldots, x_k)$, together with $\|Tx\| \leqslant \|T(|x|)\|$.

Infinite series

We show next how 1-summing operators can be characterized in terms of their actions on infinite series. This description provided the original motivation for the study of such operators.

A sequence (x_n) in X is said to be summable (or the series Σx_n to be "unconditionally Cauchy") if, given $\varepsilon > 0$, there exists N such that $\mu_1(x_{N+1}, \ldots, x_p) \leqslant \varepsilon$ for all $p > N$. This is clearly equivalent to the statement that $\Sigma \lambda_n x_n$ is a Cauchy series for every bounded scalar sequence (λ_n).

The sequence (x_n) is said to be weakly summable if there exists M such that $\mu_1(x_1, \ldots, x_n) \leqslant M$ for all n. It is easily seen that this is equivalent to either of the following statements:

(a) $\sum_n |f(x_n)| \leqslant M$ for all f in U_{X^*} ,

(b) $\sum_n \lambda_n x_n$ is a Cauchy series whenever $\lambda_n \to 0$.

Also, by applying the uniform boundedness theorem to the set of all finite sums $\Sigma \lambda_i x_i$ with each $|\lambda_i| \leqslant 1$, one sees that it is equivalent to the statement that $\Sigma |f(x_n)|$ is convergent for each f in X*.

3.24. Let T be an operator. The following statements are equivalent:

(i) T is 1-summing,

(ii) $\Sigma \|Tx_n\|$ is convergent whenever (x_n) is weakly summable,

(iii) $\Sigma \|Tx_n\|$ is convergent whenever (x_n) is summable.

Proof. Clearly, (i) implies (ii) and (ii) implies (iii). To show that (iii) implies (i), suppose that T is not 1-summing. For each n, there is a finite sequence S_n such that $\mu_1(S_n) \leqslant 2^{-n}$, while $\Sigma\{\|Tx\| : x \in S_n\} \geqslant 1$. Form a sequence (x_n) by taking the elements of $S_1, S_2, \dots$ in turn. Then it is clear that (x_n) is summable and $\Sigma \|Tx_n\|$ is divergent.

Similarly, T is 2-summing if and only if $\Sigma \|Tx_n\|^2$ is convergent whenever $\Sigma |f(x_n)|^2$ is convergent for all f in X*.

The mixed summing norm $\pi_{2,1}$

A "hybrid" summing norm $\pi_{2,1}$ is defined as follows:

$$\pi_{2,1}(T) = \sup\{(\Sigma \|Tx_i\|^2)^{1/2} : \mu_1(x_1, \dots, x_k) \leqslant 1\} .$$

(Similarly, one can define $\pi_{q,p}$ for any p, q $\geqslant$ 1.)

Clearly, we have $\|T\| \leqslant \pi_{2,1}(T) \leqslant \pi_2(T)$. One verifies as in 3.1 that $\pi_{2,1}$ is a norm and that $\pi_{2,1}(AT) \leqslant \|A\| \, \pi_{2,1}(T)$, etc. As with π_p, we write $\pi_{2,1}(X)$ for $\pi_{2,1}(I_X)$.

We would not introduce $\pi_{2,1}$ if it were not for the following fact, which suggests that it is a more "natural concept" than one might suppose at first sight. Recall from 2.12 that for elements of an inner product space,

$$(\Sigma \|x_i\|^2)^{1/2} \leqslant \mu_1(x_1, \dots, x_k) .$$

In other words, we have:

3.25. For any inner product space H, $\pi_{2,1}(H) = 1$.

It follows that for all operators from or into an inner product space, $\pi_{2,1}(T) = \|T\|$.

We return to a more thorough consideration of $\pi_{2,1}$ in section 14. In the meantime, our concern with it will be mostly limited to identity operators. As in 3.24, $\pi_{2,1}(X)$ being finite is equivalent to $\Sigma \|x_n\|^2$ being convergent for every summable sequence (x_n) in X. Orlicz (1933) showed that $L_p[0,1]$ has this property for $1 \leqslant p \leqslant 2$ (we will prove the case p = 1 in section 7, in a rather stronger form). Consequently, an infinite-dimensional space X is said to have the Orlicz property if $\pi_{2,1}(X)$ is finite.

3.26 Example. Let T be the operator $\text{diag}(\alpha_1, \dots, \alpha_n)$ in $L(\ell_\infty^n)$. Exactly as in 3.5, we see that $\pi_{2,1}(T) = \pi_2(T) = (\Sigma \alpha_i^2)^{1/2}$. In particular, $\pi_{2,1}(\ell_\infty^n) = \sqrt{n}$.

Historical note

Orlicz's theorem can be regarded as the historical beginning of the study of summing operators. It was natural to ask next whether, in particular infinite-dimensional spaces X, the series $\Sigma \|x_n\|$ is convergent for every summable sequence (x_n). Macphail (1947) recognized that this was equivalent to asking whether (in our notation) there was an upper bound to $\pi_1(E)$ for all finite-dimensional subspaces E of X. Actually, Macphail considered the reciprocal of $\pi_1(E)$, which became known as the "Macphail constant" of E. He showed that ℓ_1 does not have the above property. Dvoretzky & Rogers (1950) then proved that the same is true for every infinite-dimensional space. We shall see in section 5 how easily this result can be derived from the theory as now developed (though admittedly we will not reproduce the full strength of the theorem as stated by Dvoretzky and Rogers).

The notion 1-summing for operators other than the identity seems to have originated in Grothendieck (1955, 1956), under the name "applications semi-integrales a droite." Grothendieck's work contained a wealth of new ideas and deep theorems, and indeed underlies a large part of the contents in this book. However, it is not easy reading ! The development of the theory of 1-summing operators (and norms) in their own right was eventually taken up by Pelczynski (1962) and Pietsch (1963). The study of p-summing operators (for other p) was then initiated by Pietsch (1967).

Further examples

We conclude this section with some miscellaneous further examples.

3.27 Example. In 2.14, we found three elements of (real) ℓ_2^2 with $\|x_i\| = 1$ and $\mu_1(x_1, x_2, x_3) = 2$. It follows that $\pi_1(\ell_2^2) \geq \frac{3}{2}$.

The exact computation of $\pi_1(\ell_2^2)$ is by no means trivial; we will eventually achieve it in section 8.

3.28 Example. Let H_3 be the subspace of ℓ_∞^4 consisting of elements x such that $\Sigma x(i) = 0$. It is elementary that H_3 is isometric to ℓ_1^3. We show that $\pi_1(H_3) = 2$. (The evaluation of $\pi_1(\ell_1^n)$ is described in section 7).

It is easy to verify that for x in H_3,

$$\|x\| \leq \frac{1}{2} \sum_{i=1}^{4} |x(i)| .$$

By 3.2, it follows that $\pi_1(H_3) \leq 2$. Let $x_1 = (1,-1,0,0)$, $x_2 = (0,0,1,-1)$. Then $\mu_1(x_1, x_2) = 1$, while $\|x_1\| + \|x_2\| = 2$.

3.29 Example. Choose $a = (a_1, \ldots, a_n)$ in $\mathbb{R}^n$, and let $u_1, \ldots, u_n$ be the elements obtained by permuting the terms of a cyclically. Let T be the operator in $L(\ell_\infty^n)$ defined by: $Te_i = u_i$. Then $\pi_2(T) = \sqrt{n}\, \|a\|_2$.

To show this, observe first that $\|Tx\| = |\Sigma\, a_{\sigma(i)} x(i)|$, where σ is some permutation of $1,2, \ldots, n$. Hence, by Schwarz's inequality,

$$\|T\|^2 \leq (\Sigma\, a_i^2)\, (\Sigma\, x(i)^2) .$$

By 3.2, it follows that $\pi_2(T)^2 \leq \Sigma\, a_i^2$.

Clearly, $\|Tu_i\| = \Sigma\, a_i^2$ for each i, and $\mu_2(u_1, \ldots, u_n) = \|a\|_2$. The stated equality follows.

3.30 Example. Let E be the operator from ℓ_1^2 to ℓ_∞^2 defined by: $Ee_1 = Ee_2 = (1,1)$. Then $E = e \otimes e$, so $\pi_1(E) = \|E\| = 1$. Define S by : $Se_1 = (1,1)$, $Se_2 = (0,1)$. Let $x_1 = (1,1)$, $x_2 = (1,-1)$. Then $\mu_1(x_1,x_2) = 2$, while $\|Sx_1\| + \|Sx_2\| = 3$. Hence we have $0 \leq S \leq E$, but $\pi_1(S) > \pi_1(E)$ (compare 3.23).

3.31 Example. Let E be the subspace of ℓ_∞^3 consisting of elements x satisfying $x(1) = x(2) + x(3)$. This is not a sublattice of ℓ_∞^3 (though it is a lattice under its own ordering !). Let J be the "identity" mapping from E to ℓ_1^3.

Clearly, J is positive. We show that $\pi_1(J) > \|J\|$ (in contrast to 3.20). It is easily shown that $\|J\| = 2$. Let $x_1 = (2,1,1)$, $x_2 = (0,1,-1)$. Then $\mu_1(x_1,x_2) = 2$, while $\|Jx_1\| + \|Jx_2\| = 6$. Hence $\pi_1(J) \geqslant 3$.

Exercise. Let $X_1, \dots ,X_n$ be finite-dimensional spaces, and let X be the product space $X_1 \times \dots \times X_n$, with norm defined by: $\|x\| = \max \|x_i\|$. Show that $\pi_1(X) = \Sigma\, \pi_1(X_i)$.

4. OTHER NUCLEAR NORMS; DUALITY WITH THE SUMMING NORMS

The definition and immediate consequences

The norm ν_1 was considered in section 1. We now introduce nuclear norms ν_p for other p (including ∞). The definition perhaps seems rather contrived at first sight, but it will soon be justified by results. In particular, these norms are in trace duality with the summing norms. As with the summing norms, we concentrate our account on the "natural" cases $p = 1,2,\infty$.

Let X,Y be normed linear spaces, and let $1 \leq p < \infty$. Let p' be the conjugate index : $\frac{1}{p} + \frac{1}{p'} = 1$.

For T in FL(X,Y), we define

$$\nu_p(T) = \inf\{ (\sum_i \|f_i\|^p)^{1/p} \mu_{p'}(y_1, \dots, y_k) : T = \sum_i f_i \otimes y_i\},$$

$$\nu_\infty(T) = \inf\{ \mu_1(y_1, \dots, y_k) : T = \sum_i f_i \otimes y_i \text{ with each } \|f_i\| \leq 1\}.$$

4.1. ν_p is a norm on FL(X,Y), and $\nu_p(T) \geq \pi_p(T)$. Further :

(i) $\nu_p(f \otimes y) = \|f\| . \|y\|$,

(ii) $\nu_p(BT) \leq \|B\| \nu_p(T)$, $\nu_p(TA) \leq \nu_p(T) \|A\|$,

(iii) if $p \leq q$, then $\nu_p(T) \geq \nu_q(T)$.

Proof. Take representations

$$S = \sum_i f_i \otimes y_i, \quad T = \sum g_j \otimes z_j$$

with the appropriate quantities approximating to $\nu_p(S)$, $\nu_p(T)$. For $p = \infty$, do this with $\|f_i\| = \|g_i\| = 1$. For $p = 2$, do it with $(\Sigma \|f_i\|^2)^{1/2} = \mu_2(y_1, \dots, y_k)$ (and similarly for T). In each case, the combined representation for S+T shows that $\nu_p(S+T) \leq \nu_p(S) + \nu_p(T)$.

Let $T = \sum_i f_i \otimes y_i$, with $\mu_{p'}(y_1, \dots, y_k) = 1$. Then

$$\|Tx\| = \|\Sigma f_i(x)y_i\| \leqslant (\Sigma |f_i(x)|^p)^{1/p} .$$

It follows that $\pi_p(T) \leqslant (\Sigma \|f_i\|^p)^{1/p}$, and hence $\pi_p(T) \leqslant \nu_p(T)$ (also $\|T\| \leqslant \nu_\infty(T)$).

Statement (i) now follows, and statement (ii) is easy.

We prove (iii) in the form $\nu_\infty(T) \leqslant \nu_2(T) \leqslant \nu_1(T)$. First, let $T = \Sigma f_i \otimes y_i$ with $\|f_i\| = \|y_i\|$ and $\Sigma \|f_i\|^2 \leqslant (1+\varepsilon)\, \nu_1(T)$. Since $\mu_2(y_1, \dots, y_k) \leqslant (\Sigma \|y_i\|^2)^{1/2}$, this gives at once $\nu_2(T) \leqslant \Sigma \|f_i\|^2$.

Now let $T = \Sigma f_i \otimes y_i$ with $\mu_2(y_1, \dots, y_k) = 1$ and $\Sigma \|f_i\|^2 \leqslant (1+\varepsilon)^2 \nu_2(T)^2$. Let $\alpha_i = \|f_i\|$, $g_i = \alpha_i^{-1} f_i$, $z_i = \alpha_i y_i$. Then $T = \Sigma g_i \otimes z_i$, and we have (as in 3.3),

$$\mu_1(z_1, \dots, z_k) \leqslant (\Sigma \alpha_i^2)^{1/2} \leqslant (1+\varepsilon)\, \nu_2(T) .$$

As with ν_1, the following comments apply :

(1) If T_1 is a restriction of T, then $\nu_p(T_1) \leqslant \nu_p(T)$.

(2) Extensions (with almost the same values of ν_p) are given by taking Hahn-Banach extensions of the f_i.

(3) One must distinguish between an element of FL(X,Y) and the same mapping regarded as an element of $FL(X,Y_1)$, where Y_1 is a subspace of Y.

(4) By 1.2, we have $\nu_p(T^{**}) \leqslant \nu_p(T)$.

The study of the norms ν_p and the corresponding class of p-nuclear operators (not restricted to finite rank) was initiated by Persson & Pietsch (1969).

Trace duality with the summing norms

The definitions of the summing and nuclear norms are ready made to fit together, as follows :

<u>4.2</u> Let $S \in FL(X,Y)$ and $T \in L(Y,Z)$. Then :

(i) $\nu_1(TS) \leqslant \pi_1(T)\, \nu_\infty(S)$,

(ii) $\nu_1(TS) \leqslant \pi_2(T)\, \nu_2(S)$.

Proof. (i) Take $\varepsilon > 0$. Let $S = \sum_i f_i \otimes y_i$, with $\|f_i\| = 1$ and $\mu_1(y_1, \dots, y_k) \leqslant (1+\varepsilon)\nu_\infty(S)$. Then $TS = \sum_i f_i \otimes (Ty_i)$, and

$$\Sigma \|Ty_i\| \leq \pi_1(T)\mu_1(y_1, \dots, y_k)$$
$$\leq (1+\varepsilon)\ \pi_1(T)\nu_\infty(S)\ .$$

The proof of (ii) is similar. We take $\Sigma \|f_i\|^2 = 1$ and $\mu_2(y_1, \dots, y_k) \leq (1+\varepsilon)\ \nu_2(S)$.

In the same way, one can show that $\nu_1(TS) \leq \pi_{p'}(T)\nu_p(S)$.

4.3 Proposition. Under trace duality, $\nu_\infty^* = \pi_1$ and $\nu_2^* = \pi_2$ (and $\nu_p^* = \pi_{p'}$) . In other words, for T in $P_1(Y,X)$,

$$\pi_1(T) = \sup \{|\mathrm{trace}\ (TS)| : S \in FL(X,Y),\ \nu_\infty(S) \leq 1\}\ ,$$

and if X is finite-dimensional, then the Banach space dual of $[L(X,Y),\nu_\infty]$ identifies with $[L(Y,X),\ \pi_1]$ (and similarly for π_2,ν_2).

Proof. Take T in L(Y,X). If $\nu_\infty(S) \leq 1$, then by 4.2,

$$|\mathrm{trace}\ (TS)| \leq \nu_1(TS) \leq \pi_1(T)\ \nu_\infty(S) \leq \pi_1(T)\ .$$

Take $\varepsilon > 0$. There exist elements y_i with $\mu_1(y_1, \dots, y_k) = 1$ and $\Sigma \|Ty_i\| \geq (1-\varepsilon)\ \pi_1(T)$. Take $f_i \in X^*$ with $\|f_i\| = 1$ and $f_i(Ty_i) = \|Ty_i\|$. Let $S = \Sigma f_i \otimes y_i$. Then $\nu_\infty(S) \leq 1$, and

$$\mathrm{trace}\ (TS) = \Sigma f_i(Ty_i) \geq (1-\varepsilon)\pi_1(T)\ .$$

For the pair ν_2, π_2 , we reason similarly, choosing f_i with $\|f_i\| = \|Ty_i\|$ and $f_i(Ty_i) = \|Ty_i\|^2$. This gives S with $\nu_2(S) \leq \pi_2(T)$ and trace $(TS) \geq (1+\varepsilon)^2\ \pi_2(T)^2$.

4.4 Corollary. If X is finite-dimensional (or reflexive) and S is in FL(X,Y), then there exists T in L(Y,X) with $\pi_1(T) = 1$ and trace $(TS) = \nu_\infty(S)$. Similarly for π_2, ν_2 .

Using the corresponding duality result for ν_1 itself, we can now interchange $\pi_{p'}$ and ν_p in 4.2 (a fact that is not obvious a priori):

4.5 Let X be finite-dimensional, $S \in L(X,Y)$ and $T \in FL(Y,Z)$. Then

$$\nu_1(TS) \leq \nu_\infty(T)\,\pi_1(S), \qquad \nu_1(TS) \leq \nu_2(T)\,\pi_2(S)\ .$$

Proof. By 1.12, there is an A in L(Z,X) with $\|A\| = 1$ and

$$\begin{aligned} \nu_1(TS) &= \text{trace}\,(A.TS) \\ &= \text{trace}\,(AT.S) \\ &= \text{trace}\,(S.AT) \\ &\le \pi_1(S)\ \nu_\infty(AT) \qquad \text{(by 4.2)} \\ &\le \pi_1(S)\ \nu_\infty(T)\ . \end{aligned}$$

We defer to section 16 the question of identifying the duals of the summing norms in the sense that we have defined.

Write $\nu_p^{(n)}(T)$ for the quantity defined like $\nu_p(T)$, but only allowing representations of length not greater than n (where n, of course, is not less than the rank of T). We saw in 1.17 that this can differ from $\nu_p(T)$, and in fact $\nu_p^{(n)}$ is not a norm. However, the reasoning of 4.2 and 4.3 still holds to show that $\pi_{p'}^{(n)}$ is "dual" to $\nu_p^{(n)}$, and this sometimes yields useful facts about $\pi_{p'}^{(n)}$.

Special properties of ν_∞

4.6. If T maps from or into ℓ_∞^n, then

$$\nu_\infty(T) = \nu_\infty^{(n)}(T) = \|T\|\ .$$

Proof. First, suppose that T maps from ℓ_∞^n. Then $T = \Sigma\ x(i)(Te_i)$, and $\mu_1(Te_1, \ldots, Te_n) \le \|T\|$.

Now suppose that T maps into ℓ_∞^n. Let $Tx = \Sigma\ f_i(x)e_i$, so that $T = \Sigma\ f_i \otimes e_i$. Then $\|f_i\| \le \|T\|$ for each i, and $\mu_1(e_1, \ldots, e_n) = 1$.

From this and 4.3, we deduce at once :

4.7. If X is finite-dimensional and T is in $L(X, \ell_\infty^n)$, then $\nu_1(T) = \pi_1(T)\ (= \pi_1^{(n)}(T))$.

Recall (3.14) that the same applies to operators on ℓ_∞^n.

The last two results provide a very nice illustration of the dependence of ν_1 and ν_∞ on the range space. Let Y be a subspace ℓ_∞^N. Let J be the inclusion $Y \to \ell_\infty^N$, and let T be an operator from X to Y. Then JT is the same mapping, regarded as an operator from X to ℓ_∞^N, and

$$\nu_1(JT) = \pi_1(JT) = \pi_1(T),$$

$$\nu_\infty(JT) = \|JT\| = \|T\|\ .$$

So if $v_1(T)$ is different from $\pi_1(T)$, then it is different from $v_1(JT)$ (and similarly if $v_\infty(T) \neq \|T\|$). As we shall see, we only need to take $T = I_Y$ (for suitable Y) to obtain inequality in both cases.

The concept of an $\mathcal{L}_\infty$-space (see section 0) is the key to a painless extension of 4.6 and 4.7 to the infinite-dimensional case.

<u>4.8</u> Let X be finite-dimensional, Y an an $\mathcal{L}_\infty$-space and T an element of L(X,Y). Then $v_\infty(T) = \|T\|$, $v_1(T) = \pi_1(T)$.

Proof. Choose $\varepsilon>0$. There is a subspace Y_0 of Y, containing T(X), such that $d(Y_0, \ell_\infty^N) \leqslant 1 + \varepsilon$ for some N. Let T_0 be the same mapping T, regarded as an element of $L(X,Y_0)$. Then $v_1(T) \leqslant v_1(T_0)$, while $\pi_1(T) = \pi_1(T_0)$. It follows easily from 4.7 that $v_1(T_0) \leqslant (1+\varepsilon)\,\pi_1(T_0)$. Hence $v_1(T) \leqslant (1+\varepsilon)\,\pi_1(T)$. Similarly for v_∞.

Using 4.6, we now derive two equivalent formulations, both very "natural", of the meaning of v_∞.

<u>4.9.</u> Let X be finite-dimensional, and let T be in L(X,Y). Then $v_\infty(T)$ is the infimum of the numbers α for which the following statement holds : given any space Z containing X, there is an extension $\hat{T} : Z \to Y$ of T with $\|T\| \leqslant \alpha$.

Proof. Firstly, for any $\varepsilon>0$, there is an extension $\hat{T}$ with $\|\hat{T}\| \leqslant v_\infty(\hat{T}) \leqslant (1+\varepsilon)\, v_\infty(T)$.

By 0.14, there is a space Z containing X (strictly, an isometric copy of X) with $d(Z, \ell_\infty^N) \leqslant 1+\varepsilon$ for suitable N. Suppose that T has an extension with $\|\hat{T}\| \leqslant \alpha$. By 4.6, $v_\infty(\hat{T}) \leqslant \alpha(1+\varepsilon)$. The statement follows.

<u>4.10 Corollary.</u> If X is finite-dimensional, then $v_\infty(I_X) = \lambda(X)$ (the projection constant of X).

<u>4.11.</u> Let T be in FL(X,Y). Then $v_\infty(T)$ is the infimum of $\|T_2\|.\|T_1\|$ taken over all factorizations $T = T_2T_1$, where $T_1 \in L(X, \ell_\infty^k)$, $T_2 \in L(\ell_\infty^k, Y)$ and k is any integer.

Proof. Given such a factorization, we have by 4.6 :

$$v_\infty(T) \leqslant v_\infty(T_2)\,\|T_1\| = \|T_2\|.\|T_1\|\,.$$

For the converse, take $\varepsilon > 0$. Let $T = \sum_i f_i \otimes y_i$, where $\|f_i\| = 1$ for each i and $\mu_1(y_1, \dots, y_k) \leq (1+\varepsilon)\, \nu_\infty(T)$. Define $T_1 : X \to \ell_\infty^k$ and $T_2 : \ell_\infty^k \to Y$ by :

$$T_1 x = \Sigma\, f_i(x) e_i \, ,$$

$$T_2 e_j = y_i \, .$$

Then $T = T_2 T_1$, $\|T_1\| = 1$ and by 2.5, $\|T_2\| \leq (1+\varepsilon)\, \nu_\infty(T)$.

Note. For mappings between finite-dimensional spaces, it follows that $\nu_\infty(T^*)$ equates to the quantity defined in the same way by factorizing T through ℓ_1^k . Of course, $\nu_\infty(T^*)$ is obtained by interchanging $\| \ \|$ and μ_1 in the definition of $\nu_\infty(T)$. In general, it bears no relation to $\nu_\infty(T)$. For example, if T is the identity in ℓ_∞^n, then T^* is the identity in ℓ_1^n , and $\nu_\infty(T) = 1$, while $\nu_\infty(T^*) = \lambda(\ell_1^n)$. (But of course $\nu_\infty(T^{**}) = \nu_\infty(T)$ when the spaces are finite-dimensional).

Exercise. Show that for a diagonal operator D from ℓ_∞^k to ℓ_p^k, $\nu_p(D) = \|D\|$. Deduce that for T in FL(X,Y), $\nu_p(T)$ is the infirmum of $\|T_2\| . \|D\| . \|T_1\|$ taken over factorizations $T = T_2 D T_1$, where $T_1 \in L(X, \ell_\infty^k)$, $T_2 \in L(\ell_p^k, Y)$ and D is a diagonal operator from ℓ_∞^k to ℓ_p^k .

For identity mappings, $\nu_\infty^{(n)}$ has a nice interpretation.

4.12. Let dim X = n. Then

$$\nu_\infty^{(n)}(I_X) = d(X, \ell_\infty^n) \, .$$

Proof. Representations of length n for I_X are of the form $\Sigma\, f_i \otimes b_i$, where (b_i) is a base of X, and (f_i) is the biorthogonal set of functionals.

Given a base (b_i) of X, define an isomorphism $\ell_\infty^n \to X$ by: $Te_i = b_i$. Then $\|T\| = \mu_1(b_1, \dots, b_n)$. Further, $T^{-1}x = \Sigma\, f_i(x) e_i$, so $\|T^{-1}\| = \max \|f_i\|$. All isomorphisms are of this form. The statement follows.

Compare this with the equality $\nu_\infty(I_X) = \lambda(X)$. Of course, $\lambda(X) \leq d(X, \ell_\infty^n)$. It is a long unsolved question whether there is a constant C (independent of n) such that $d(X, \ell_\infty^n) \leq C\lambda(X)$ for all n-dimensional spaces X. We give a partial result in this direction later (12.11).

Special properties of ν_2

The outstanding special property of ν_2 is that it coincides with π_2. This will be proved in section 5. For the moment, we just remark that we have already (in 3.9) essentially proved this statement for operators between Hilbert spaces : with the notation used there,

$$T = \sum_j (T^*f_j) \otimes f_j ,$$

so $$\nu_2(T)^2 \leqslant \sum_j \|T^*f_j\|^2 = \pi_2(T)^2 .$$

Exercise. Prove by the steps indicated that the value of ν_2 is independent of the range space. Given $\mu_2(y_1, \dots ,y_k) = 1$, let Z be a subspace of $Y_0 = \mathrm{lin}(y_1, \dots ,y_k)$. By considering the mappings $A : \ell_2^k \to Y_0$ given by $Ae_i = y_i$, show that there is a projection P of Y_0 onto Z with $\mu_2(Py_1, \dots ,Py_k) \leqslant 1$.

"Lifting" and right inverses

As with the extension problem, the solution to the "lifting" problem is effectively built in to the definition of ν_p - though we have to consider $\nu_p(T^*)$ rather than $\nu_p(T)$ (as we know, for p = 1 these coincide). Recall that we say that Q is an "M-open" operator of Y onto Z if for each $z \in Z$, there exists $y \in Y$ with $Qy = z$ and $\|y\| \leqslant M\|z\|$.

4.13. Let X,Y,Z be normed linear spaces (Z finite-dimensional). Let Q be an M-open operator of Y onto Z. Given T in L(X,Z) and $\varepsilon>0$, there exists T_1 in FL(X,Y) such that $T = QT_1$ and $\nu_p(T_1^*) \leqslant (1+\varepsilon)M\, \nu_p(T^*)$.

Proof. Express T as $\Sigma\, f_r \otimes z_r$, with $\mu_{p'}(z_1, \dots ,z_k) \leqslant (1+\varepsilon)\nu_p(T^*)$ and $\Sigma\, \|z_r\|^p \leqslant 1$ (for $p = \infty$, $\|z_r\| \leqslant 1$ for each r). Let y_r be such that $Qy_r = z_r$ and $\|y_r\| \leqslant M\|z_r\|$, and let $T_1 = \Sigma\, f_r \otimes y_r$.

The special case $Z = X$, $T = I_X$ gives :

4.14 Corollary. Let X be finite-dimensional, and let Q be an M-open operator of Y onto X. Then, for any $\varepsilon > 0$, there is an operator J in L(X,Y) such that $QJ = I_X$ and $\nu_p(J^*) \leqslant (1 + \varepsilon)M\, \nu_p(I_{X^*})$. Then JQ is a projection of Y onto J(X).

For a subspace E of Y, we denote by E^0 the annihilator of E in Y^*. If E is of codimension n (that is, it has an n-dimensional complement), then $\dim E^0 = n$.

<u>4.15 Corollary.</u> Let E be a closed, n-codimensional subspace of Y (which may be infinite-dimensional). Then, given $\varepsilon > 0$, there exists a projection P on Y with *kernel* E and $\|P\| \leqslant (1 + \varepsilon)\, \lambda(E^0)$.

Proof. In 4.4, let Q be the quotient map of Y onto Y/E , and let $P = JQ$. Then $\ker P = \ker Q = E$. Also, $(Y/E)^*$ is isometric to E^0. Apply 4.14 with $p = \infty$: we have $\|J\| \leqslant \nu_\infty(J^*)$ and $\nu_\infty(I_{E^0}) = \lambda(E^0)$.

5. PIETSCH'S THEOREM AND ITS APPLICATIONS

The theorem

We now come to the fundamental theorem on p-summing operators. We have seen (3.17) that if T is an operator defined on a subspace X of $\ell_\infty(S)$ and if there is a functional ϕ on $\ell_\infty(S)$ such that $\|Tx\|^p \leqslant \phi(|x|^p)$ for all $x \in X$, then $\pi_p(T)^p \leqslant \|\phi\|$. Pietsch's theorem states that, conversely, such a functional ϕ always exists. Its application is quite general since, of course, every Banach space can be embedded in a suitable $\ell_\infty(S)$ (note that there is no requirement that $|x|^p$ belongs to X whenever x does). For $p = 1$, Pietsch published the theorem in a slightly disguised form in (1961), and in more or less the present form in (1963). The version for general p appeared in (1967).

First, we mention one special case in which the proof is easy. Suppose that T is defined on ℓ_∞^n. Then $\pi_1(T) = \Sigma\, \alpha_i$, where $\alpha_i = \|Te_i\|$ (3.14). Now $Tx = \Sigma\, x(i)(Te_i)$, so $\|Tx\| \leqslant \Sigma\, \alpha_i|x(i)|$. This is $\phi(|x|)$, where $\|\phi\| = \Sigma\, \alpha_i$.

For the general case, we use the following lemma of F.F. Bonsall. We say that a real-valued function q is "superlinear" if -q is sublinear. A "wedge" is a subset admitting addition and positive scalar multiplication.

5.1 Lemma. Let Q be a wedge in a (real) linear space X. Suppose that p is sublinear on X and q is superlinear on Q, with $q(y) \leqslant p(y)$ for all $y \in Q$. Then there is a linear functional f on X such that

$$f(x) \leqslant p(x) \qquad \text{for all } x \in X,$$
$$f(y) \geqslant q(y) \qquad \text{for all } y \in Q.$$

Proof. For x in X, define

$$r(x) = \inf\{p(x+y) - q(y) : y \in Q\}.$$

Since $p(x+y) + p(-x) \geqslant p(y) \geqslant q(y)$, we have $r(x) \geqslant -p(-x)$. It is elementary that r is sublinear, and clearly $r(x) \leqslant p(x)$ for $x \in X$, while $r(-y) \leqslant -q(y)$ for $y \in Q$. By the Hahn-Banach theorem, there is a linear functional f on X with $f(x) \leqslant r(x)$ for all x. Then f has the required properties.

<u>5.2 Theorem.</u> Let X be a linear subspace of $\ell_\infty(S)$. Let T be a p-summing operator defined on X (for any $p \geqslant 1$). Then there is a positive linear functional ϕ on $\ell_\infty(S)$ such that $\|\phi\| = \pi_p(T)^p$ and

$$\|Tx\|^p \leqslant \phi(|x|^p) \qquad \text{for } x \in X .$$

Proof. Consider the case of real scalars first. Let Q be the set of non-negative functions in $\ell_\infty(S)$. Given $y \in Q$, let $V(y)$ be the set of finite sequences $(x_1, \dots, x_k)$ in X with $\Sigma |x_i|^p \leqslant y$. Let

$$q(y) = \sup \{\Sigma \|Tx_i\|^p : (x_1, \dots, x_k) \in V(y)\} .$$

For $(x_1, \dots, x_k)$ in $V(y)$, we have by 2.6 ,

$$\mu_p(x_1, \dots, x_k)^p = \|\sum_i |x_i|^p\| \leqslant \|y\| ,$$

and hence

$$\sum_i \|Tx_i\|^p \leqslant \rho\|y\| ,$$

where $\rho = \pi_p(T)^p$. Therefore $q(y) \leqslant \rho\|y\|$. It is elementary that q is superlinear, and clearly for $x \in X$, we have $\|Tx\|^p \leqslant q(|x|^p)$.

By the lemma, there is a functional ϕ on $\ell_\infty(S)$ such that $\phi(z) \leqslant \rho\|z\|$ for all z in $\ell_\infty(S)$ (so $\|\phi\| \leqslant \rho$), and $\phi(y) \geqslant q(y)$ ($\geqslant 0$) for $y \in Q$. In particular, $\phi(|x|^p) \geqslant \|Tx\|^p$ for $x \in X$.

In the complex case, let $\ell_\infty(S)$, $\ell_\infty^C(S)$ denote the spaces of bounded real and complex functions on S respectively. The above proof gives a positive real functional ϕ on $\ell_\infty(S)$ satisfying the stated conditions (it doesn't matter that X is not contained in $\ell_\infty(S)$!). There is an obvious way to extend ϕ to give a (complex) linear functional on $\ell_\infty^C(S)$: let $\phi(x+iy) = \phi(x) + i\phi(y)$. It is elementary that this does not increase $\|\phi\|$.

<u>Remarks</u> (1) Suppose that X is embedded in C(K) (e.g. with $K = U_{X^*}$). Then there is a positive functional ϕ on C(K) as stated in the theorem. By the Riesz representation theorem, this functional can be identified with a Borel measure on K. The theorem is often quoted in this form. However, this is to obscure the fact that the theorem is simply a statement about the existence of a functional, not a statement about measures. For some

applications (notably the extension theorem) it is important that X can be embedded in <u>any</u> space of the form $\ell_\infty(S)$ or C(K), not only in $C(U_{X^*})$.

(2) A sketch of an alternative proof is as follows. Assume $\rho = 1$. Let

$$F_1 = \{y \in \ell_\infty(S) : \sup y(s) < 1\},$$

$$F_2 = \mathrm{co}\{|x|^p : x \in X \text{ and } \|Tx\| = 1\}.$$

One verifies that F_1 and F_2 are disjoint. The separation theorem then gives the required functional.

Given a positive functional ϕ on $\ell_\infty(S)$, one obtains a semi-inner-product by putting $\langle x,y\rangle = \phi(xy)$ (real case) or $\phi(x\bar{y})$ (complex case). Then $\phi(|x|^2) = \langle x,x\rangle$. This is the key to a number of important applications specific to the case p = 2, obtained by taking advantage of the nice properties of Hilbert spaces.

<u>Exercise.</u> Let $f_1, \ldots, f_n$ be functionals on C(K). Write $\Sigma \|f_i\|^2 = M^2$. Show that there is a positive functional ϕ on C(K) such that $\Sigma f_i(x)^2 \leqslant \phi(x^2)$ for all x and $\|\phi\| = M^2$. Is this obvious without Pietsch's theorem ?

<u>Exercise.</u> Adapt the proof of 5.2 to obtain the following variant for lattices. Let T be a 1-summing operator on a normed lattice X. Then there is a positive functional ϕ on X such that $\|\phi\| = \pi_1(T)$ and $\|Tx\| \leqslant \phi(|x|)$ for all $x \in X$.

(However, the converse is not true. This will be clarified in section 18).

Operators on finite-dimensional spaces.

As a first application of Pietsch's theorem, we obtain the promised converse of 3.2.

<u>5.3 Proposition.</u> Let T be an operator on a finite-dimensional space X, and let $\varepsilon > 0$, $p \geqslant 1$. Then there exist elements $f_1, \ldots, f_k$ of X^* such that $\|Tx\|^p \leqslant \sum_j |f_j(x)|^p$ for all $x \in X$, and

$$(\Sigma \|f_j\|^p)^{1/p} \leqslant (1+\varepsilon)\, \pi_p(T).$$

Proof. It is sufficient to prove the statment (without ε) for the case when X is a subspace of some ℓ_∞^k, since every X is almost isometric to such a subspace (see 0.13).

Let ϕ be the functional given by Pietsch's theorem. We can express ϕ in the form

$$\phi(y) = \sum_j \alpha_j^p y(j) ,$$

where $\|\phi\| = \Sigma \alpha_j^p = \pi_p(T)^p$. Then $\|Tx\|^p \leq \Sigma |f_j(x)|^p$ for $x \in X$, where $f_j(x) = \alpha_j x(j)$. Clearly, $\|f_j\| \leq \alpha_j$.

<u>Notes.</u> (1) The number of functionals in 5.3 is bounded by the k for which X is embedded in ℓ_∞^k . It is easy to see that this number may have to exceed the rank of T. Consider the identity in an n-dimensional space X. We show later that there are cases for which $\pi_1(I_X) \leq \sqrt{2n}$. Suppose we have n functionals f_i such that $\|x\| \leq \Sigma |f_i(x)|$ for all x. Then the f_i are linearly independent, so there exist $b_j \in X$ such that $f_i(b_j) = \delta_{ij}$. From our hypothesis, $\|b_i\| \leq 1$, so $\|f_i\| \geq 1$, and hence $\Sigma \|f_i\| \geq n$.

(2) Let K be a norming subset of U_{X^*}. The near-isometric embedding into ℓ_∞^k is of the form

$$Ax = [h_1(x), \dots ,h_k(x)]$$

where the h_j are in K. From the way that co-ordinate functionals were used in the proof, it is clear that the f_j in 5.3 can be taken to be scalar multiples of elements of K.

We have seen very easily (3.14, 4.6) that for operators defined on ℓ_∞^n , $\nu_1(T) = \pi_1(T)$ and $\nu_\infty(T) = \|T\|$. We can now show that the same is true for general p (and in particular, p = 2).

<u>5.4 Proposition.</u> For any operator T on ℓ_∞^n, we have $\nu_p(T) = \pi_p(T)$ for all p.

Proof. We assume $1 < p < \infty$. Let ϕ be the functional given by Pietsch's theorem, so that

$$\phi(x) = \sum_j \alpha_j^p x(j) ,$$

where $\|\phi\| = \Sigma \alpha_j^p = \pi_p(T)^p$. Let K be the set of j for which $\alpha_j > 0$. For $j \in K$, let $f_j(x) = \alpha_j x(j)$ and $u_j = \alpha_j^{-1}(Te_j)$. For other j, let $f_j = u_j = 0$. Then

$$Tx = \sum_j x(j)\,(Te_j) = \sum_j f_j(x)u_j .$$

Clearly, $\Sigma \|f_j\|^p = \Sigma \alpha_j^p = \pi_p(T)^p$. The result follows if we can show that $\mu_{p'}(u_1, \ldots, u_k) \leq 1$. Take scalars λ_j with $\Sigma |\lambda_j|^p \leq 1$. Then $\Sigma \lambda_j u_j = Tx$, where $x(j) = \alpha_j^{-1}\lambda_j$ for $j \in K$ (and 0 otherwise). Hence

$$\|\Sigma \lambda_j u_j\|^p = \|Tx\|^p \leq \sum_j \alpha_j^p |x(j)|^p = \Sigma |\lambda_j|^p \leq 1 ,$$

which proves the required statement.

As in 4.7, we have at once by trace duality :

5.5 Corollary. If X is finite-dimensional and T is in $L(X, \ell_\infty^n)$, then $\nu_p(T) = \pi_p(T) = \pi_p^{(n)}(T)$ for all p.

For the case $p = 2$, we will see below that the same statement holds for operators on *any* finite-dimensional space.

An application to finite-dimensional spaces

Before continuing with the general discussion of Pietsch's theorem, we describe a beautiful application of our results on 2-summing norms.

5.6 Theorem. Let $(X, \|\ \|)$ be an n-dimensional normed linear space (real or complex). Then :

(i) there is an inner product on X giving a norm $|\ |_\phi$ such that $\|x\| \leq |x|_\phi \leq \sqrt{n}\,\|x\|$ for all $x \in X$; hence $d(X, \ell_2^n) \leq \sqrt{n}$;

(ii) $\lambda(X) \leq \sqrt{n}$.

Proof. We give the details for the real case. The complex case requires routine minor modifications.

(i) Embed X in a space $\ell_\infty(S)$. By 3.8, $\pi_2(I_X) \leq \sqrt{n}$. So by Pietsch's theorem, there is a positive linear functional ϕ on $\ell_\infty(S)$ such that $\|\phi\| \leq n$ and $\|x\|^2 \leq \phi(x^2)$ for all $x \in X$. Define a semi-inner-product on $\ell_\infty(S)$ by putting $\langle x,y\rangle = \phi(xy)$. The corresponding seminorm $|\ |_\phi$ satisfies

$$|y|_\phi^2 = \phi(y^2) \leq n\|y^2\| = n\|y\|^2$$

for all $y \in \ell_\infty(S)$. Also, $\|x\|^2 \leq \phi(x^2) = |x|_\phi^2$ for all $x \in X$ (so $|\ |_\phi$ is a norm, not just a seminorm, on X).

(ii) Let P be the orthogonal projection of $\ell_\infty(S)$ onto X with respect to $|\ |_\phi$. (The fact that $|\ |_\phi$ is only a seminorm on $\ell_\infty(S)$ doesn't matter : $Py = \Sigma \langle y,b_i \rangle b_i$, where (b_i) is an orthonormal base of X). Then

$$\|Py\| \leqslant |Py|_\phi \leqslant |y|_\phi \leqslant \sqrt{n}\ \|y\| ,$$

so $\|P\| \leqslant \sqrt{n}$. (In fact, the inequality $\|Py\|^2 \leqslant \phi(y^2)$ shows that $\pi_2(P) \leqslant \sqrt{n}$.)

5.7 Corollary. If X,Y are any two n-dimensional normed linear spaces, then $d(X,Y) \leqslant n$.

Notes (1) Statement (i) in 5.6 was originally proved by F. John (1948), by much harder methods. Statement (ii) was derived from John's result by Kadec & Snobar (1971).

(2) It is elementary that $d(\ell_1^n, \ell_2^n) = \sqrt{n}$ (see 6.10), so the $\sqrt{n}$ in (i) is best possible. Also, we will show in section 6 that ℓ_1^n and ℓ_2^n have projection constants not less than $\sqrt{n/2}$ (though not actually equal to $\sqrt{n}$).

(3) Gluskin (1981) has shown that there exists C such that for each n, there are n-dimensional spaces X,Y with $d(X,Y) \geqslant Cn$. As mentioned earlier, it is not known whether $d(X, \ell_\infty^n) \leqslant C\ \lambda(X)$ (or $\leqslant C\sqrt{n}$) for all n-dimensional X.

(4) By 4.15, it follows that if E is a closed, n-codimensional subspace of Y (not necessarily finite-dimensional), then there is a projection P with *kernel* E and $\|P\| \leqslant (1+\varepsilon)\sqrt{n}$.

Factorization and extension

In 5.6, we have seen one example of what can be done (in the case p = 2) using the inner product derived from Pietsch's functional. We now show how the same idea leads to a factorization theorem for 2-summing operators in general. This result is also known as Pietsch's theorem, though a version was published by Pelczynski (1962), and the idea is present in Grothendieck (1956). At the same time, we obtain a "Hahn-Banach" theorem for operators : any 2-summing operator can be extended without increasing the value of π_2.

Let ϕ be a positive functional on $\ell_\infty(S)$. As before, a semi-inner-product is defined on $\ell_\infty(S)$ by: $\langle x,y \rangle = \phi(xy)$ (or $\phi(x\bar{y})$ in the complex case). Let $|\ |_\phi$ be the associated seminorm. The quotient by

$\{x : |x|_\phi = 0\}$ is an inner product space (in the case of ℓ_∞^n, this just amounts to leaving out some of the co-ordinates). The completion of this inner product space is a Hilbert space, which we denote by $L_2(\phi)$. Note that this construction has nothing to do with measure theory ! By a slight abuse, we continue to use the notation $\langle\ \rangle$ and $|\ |_\phi$ in $L_2(\phi)$. Let J be the natural mapping from $\ell_\infty(S)$ to $L_2(\phi)$: strictly, this is the formal identity followed by the quotient mapping. Then $|Jx|_\phi^2 = \phi(x^2)$, so (by 3.17) we have $\pi_2(J)^2 \leqslant \|\phi\|$. We are now ready to formulate the factorization theorem :

5.8 Theorem. Let X,Y be normed linear spaces (Y complete), and let T be a 2-summing operator from X to Y. Then there exist a Hilbert space H and operators $T_1 : X \to H$ and $T_2 : H \to Y$ such that $T = T_2T_1$ and $\pi_2(T_1) = \pi_2(T)$, $\|T_2\| = 1$.

Proof. Embed X in some $\ell_\infty(S)$. Let ϕ be the functional given by Pietsch's theorem, and let $L_2(\phi)$ and J be as above. Clearly, $\pi_2(J) \leqslant \pi_2(T)$. Let $T_1 = J|_X$, and let H be the closure of J(X) in $L_2(\phi)$. For $x \in X$, we have

$$\|Tx\|^2 \leqslant \phi(x^2) = |T_1x|_\phi^2 .$$

In particular, if $T_1x = 0$, then $Tx = 0$, so we can define T_2 on $T_1(X)$ by: $T_2(T_1x) = Tx$. Further, $\|T_2(T_1x)\| \leqslant |T_1x|_\phi$, so $\|T_2\| \leqslant 1$. Extend T_2 by continuity to the domain H.

If Y is incomplete, we still obtain a factorization as above, with H a (possibly incomplete) inner product space.

We now make use of the existence of orthogonal projections in Hilbert spaces to derive the stated extension theorem.

5.9 Theorem. Let X be a subspace of a normed linear space X_1, and let Y be a Banach space. Let T be a 2-summing operator from X to Y. Then T has an extension $T_1 : X_1 \to Y$ with $\pi_2(T_1) = \pi_2(T)$.

Proof. Embed X_1 in a space $\ell_\infty(S)$. Continue to use the above notation. Let P be the orthogonal projection of $L_2(\phi)$ onto H, and let $T_1 = T_2PJ$. Then T_1 extends T and

$$\pi_2(T_1) \leqslant \|T_2P\|\ \pi_2(J) \leqslant \pi_2(T) .$$

Exercise. Use orthogonal projections to show that the operator T_2 in 5.8 can be taken to be one-to-one.

How much of the above construction applies to other p, in particular $p = 1$? A seminorm $\| \ \|_1$ is defined by : $\|x\|_1 = \phi(|x|)$. By taking the quotient with $\{x : \|x\|_1 = 0\}$ and then the completion, we obtain a space which we denote by $L_1(\phi)$. One can show that this is indeed an $\mathcal{L}_1$-space. For a 1-summing operator, the method of 5.8 gives a factorization through a subspace of $L_1(\phi)$. However, this subspace need not be complemented or an $\mathcal{L}_1$-space.

A simple example is enough to show that 1-summing operators cannot be extended with preservation of π_1 (although, of course, they must have 2-summing extensions !).

5.10 Example. Let E be a subspace of ℓ_∞^N isometric to ℓ_1^n (there is such a subspace with $N = 2^n$). We show later (7.12) that $\pi_1(I_E) \leq \sqrt{2n}$. Let $P : \ell_\infty^N \to E$ be an extension of I_E, in other words, a projection onto E. By 3.14 and 1.10, $\pi_1(P) = \nu_1(P) \geq n$.

The equivalence of ν_2 and π_2

The extension theorem gives at once the promised generalization of 5.4 :

5.11 Theorem. Let T be any operator defined on a finite-dimensional space. Then $\nu_2(T) = \pi_2(T)$.

Proof. Let T be in L(X,Y). As usual, it is sufficient to consider the case where X is a subspace of some ℓ_∞^k. There is an extension $T_1 : \ell_\infty^k \to Y$ with $\pi_2(T_1) = \pi_2(T)$ (note that T(X) is finite-dimensional, so complete). By 5.4, $\nu_2(T_1) = \pi_2(T_1)$. The statement follows.

We will see later (section 17) that the same holds for all finite-rank operators, even when defined on infinite-dimensional spaces.

Because of 5.11, we can replace ν_2 by π_2 in the duality results of section 4. In particular, from 4.2 we have $\nu_1(TS) \leq \pi_2(T)\pi_2(S)$ for S with finite-dimensional domain. Without any finite dimensionality restrictions, we deduce :

5.12 Proposition. If S is in $P_2(X,Y)$ and T in $P_2(Y,Z)$, then TS is in $P_1(X,Z)$, and $\pi_1(TS) \leq \pi_2(T)\,\pi_2(S)$.

Proof. Let X_1 be a finite-dimensional subspace of X_1 and let $S_1 = S|_{X_1}$. Then

$$\pi_1(TS_1) \leq \nu_1(TS_1) \leq \pi_2(T)\pi_2(S_1) \leq \pi_2(T)\pi_2(S)\ .$$

The statement follows.

We are now in a position to replace the inequality in 3.8 (which has served us so well) by equality :

5.13 Proposition. For any n-dimensional normed linear space X, we have $\pi_2(I_X) = \sqrt{n}$.

Proof. Since $I_X = I_X^2$, we have

$$n = \nu_1(I_X) \leq \pi_2(I_X)^2\ .$$

Hence $\pi_1(I_X) \geq \sqrt{n}$. Further, we deduce :

5.14 Corollary. If X is infinite-dimensional, then I_X is not 2-summing. Nor is any isomorphism between infinite-dimensional spaces.

Proof. $\pi_2(I_X)$ would have to exceed $\sqrt{n}$ for all n. If an isomorphism T were 2-summing, then $T^{-1}T = I_X$ would be too.

Hence I_X (for infinite-dimensional X) is not 1-summing either. Recall that by 3.24 this means that there is a summable sequence (x_n) in X with $\Sigma\ \|x_n\|$ divergent. This is the theorem of Dvoretzky & Rogers (1950).

Further applications

We finish this section with some further applications specific to infinite dimensions. These results are not essential for our later theory, and the reader is at liberty to leave them out.

5.15. If X,Y are Banach spaces, and if there is a 2-summing operator mapping X onto Y, then Y is isomorphic to a Hilbert space.

Proof. By 5.8, Y is the image of a Hilbert space under a continuous linear mapping. Such a space is isomorphic to a Hilbert space.

We have already seen (3.12) that any 2-summing operator between Hilbert spaces is compact. By the factorization theorem, the same is true for operators from a Hilbert space to a Banach space. In the general case, we have two weaker properties:

5.16. Any 2-summing operator between Banach spaces is weakly compact.

Proof. The operator factorizes through a Hilbert space.

An operator T is "completely continuous" if $\|Tx_n\| \to 0$ whenever (x_n) is weakly convergent to 0. This is equivalent to T(K) being norm-compact whenever K is weakly compact. If X is reflexive or X^* is separable, then all completely continuous operators on X are compact. As a generalization of the above statement for Hilbert spaces, we have :

5.17. Any 2-summing operator between Banach spaces is completely continuous.

Proof. Embed X in a space C(K), and let ϕ be as in Pietsch's theorem. Let $x_n \to 0$ weakly in X. Regarded as elements of C(K), this says that $x_n(t) \to 0$ for each $t \in K$. Hence $x_n^2 \to 0$ weakly, so $\phi(x_n^2) \to 0$. It follows that $\|Tx_n\| \to 0$.

5.18 Corollary. If S and T are 2-summing, then TS is compact.

Proof. This follows from S being weakly compact and T completely continuous.

It can actually be shown that the product of two 2-summing operators is nuclear, in the sense mentioned after 1.13 (this strengthens both 5.18 and 5.12).

We saw in 3.13 that compact operators, even between Hilbert spaces, are not always 2-summing. Conversely, there are 2-summing (even 1-summing) operators that are not compact (see 6.2).

6. AVERAGING : TYPE 2 AND COTYPE 2 CONSTANTS

The basic averaging result and its applications

So far, we have been essentially concerned with general results. The emphasis will now shift to results applying to particular spaces, or to put it another way, to properties of particular spaces that are (or can be) formulated in terms of the summing and nuclear norms. The following very simple "averaging" result underlies or motivates a high proportion of this work.

Let D_n denote $\{-1,1\}^n$, the set of 2^n elements of form $\varepsilon = (\varepsilon_1, \dots, \varepsilon_n)$ with each $\varepsilon_i \in \{-1,1\}$.

We make repeated use of the obvious fact that for fixed, distinct integers i,j,

$$\sum_{\varepsilon \in D_n} \varepsilon_i \varepsilon_j = 0 .$$

Since D_n is a subset of $\mathbb{R}^n$, we can form inner products in the usual way between elements of D_n and elements of $\mathbb{R}^n$ (or $\mathbb{C}$).

<u>6.1 Proposition.</u> Let K be $\mathbb{R}$ or $\mathbb{C}$. Then :

(i) $I_{K^n} = \frac{1}{2^n} \sum_{\varepsilon \in D_n} \varepsilon \otimes \varepsilon$. That is, for $x \in K^n$,

$$x = \frac{1}{2^n} \sum_{\varepsilon \in D_n} \langle x, \varepsilon \rangle \varepsilon .$$

(ii) For $x,y \in K^n$,

$$\frac{1}{2^n} \sum_{\varepsilon \in D_n} \langle x, \varepsilon \rangle \langle \varepsilon, y \rangle = \langle x, y \rangle .$$

(iii) For $x \in K^n$,

$$\frac{1}{2^n} \sum_{\varepsilon \in D_n} |\langle x, \varepsilon \rangle|^2 = \|x\|_2^2 .$$

Proof. Component j of $\frac{1}{2^n}\sum_{\varepsilon}\langle x,\varepsilon\rangle\,\varepsilon$ is

$$\frac{1}{2^n}\sum_{\varepsilon}\varepsilon_j\sum_i \varepsilon_i x(i) = \frac{1}{2^n}\sum_i x(i)\sum_{\varepsilon}\varepsilon_i\varepsilon_j .$$

When $i \neq j$, we have $\sum_{\varepsilon}\varepsilon_i\varepsilon_j = 0$, and clearly $\frac{1}{2^n}\sum_{\varepsilon}\varepsilon_j^2 = 1$. Hence the above expression equals x(j), and (i) is proved. Statement (ii) follows on taking the inner product with y, and statement (iii) is obtained by putting $y = x$.

Recall that $I_{p,q}^{(n)}$ denotes the identity in $\mathbb{R}^n$ (or $\mathbb{C}^n$), regarded as an operator from ℓ_p^n to ℓ_q^n, and $I_{p,q}$ denotes the identity operator from ℓ_p to ℓ_q (for $p \leq q$). Proposition 6.1 enables us to evaluate $\alpha(I_{p,q}^{(n)})$ (where α is one of the summing or nuclear norms) in several of the cases not covered by our earlier results. In certain cases, this does not grow with n : this leads to a corresponding infinite-dimensional version.

6.2 Proposition. $\nu_1(I_{1,\infty}^{(n)}) = \pi_1(I_{1,\infty}^{(n)}) = 1$ for each n ; further, $\pi_1(I_{1,\infty}) = 1$.

Proof. The statement for ν_1 follows at once from 6.1 (i), since $\|\varepsilon\|_\infty = 1$ for each ε ; of course, this is the norm when ε is regarded as a functional on ℓ_1^n.

This gives $\pi_1(I_{1,\infty}^{(n)}) = 1$. Let E_n denote the obvious copy of ℓ_1^n in ℓ_1. For any operator T on ℓ_1, we have $\pi_1(T) = \sup_n \pi_1(T|_{E_n})$; hence $\pi_1(I_{1,\infty}) = 1$.

Remarks. (1) Hence also $\pi_2(I_{1,\infty}) = 1$.

(2) $I_{1,\infty}$ is not compact.

(3) Let T be the operator $\mathrm{diag}(\alpha_1, \ldots, \alpha_n)$. As before, let $T_{p,q}$ denote T regarded as an operator from ℓ_p^n to ℓ_q^n. Note that $\|T_{1,1}\| = \|T_{1,\infty}\| = \max |\alpha_i|$ (= M, say). Since $T_{1,\infty} = I_{1,\infty}T_{1,1}$, we have that $\nu_1(T_{1,\infty}) \leq \nu_1(I_{1,\infty})\|T_{1,1}\|$, and hence $\nu_1(T_{1,\infty}) = M$.

6.3 Proposition. $\pi_2(I_{1,2}^{(n)}) = 1$ for each n; $\pi_2(I_{1,2}) = 1$.

Proof. The statement for $I_{1,2}^{(n)}$ follows at once from 3.2 and 6.1(iii) (again, $\|\varepsilon\| = 1$ as a functional on ℓ_1^n). The statement for $I_{1,2}$ follows as in 6.2.

<u>6.4.</u> $\nu_1(I^{(n)}_{1,2}) = \nu_1(I^{(n)}_{2,\infty}) = \sqrt{n}$.

Proof. 6.1(i), with appropriate norms, shows that both quantities are not greater than $\sqrt{n}$. Equality for $I_{1,2}$ (temporarily, we drop the superscript (n)) follows from :

$$n = \nu_1(I_{1,1}) \leqslant \nu_1(I_{1,2})\, \|I_{2,1}\| = \sqrt{n}\, \nu_1(I_{1,2}) .$$

Similar reasoning applies to $I_{2,\infty}$.

Let $\mu_2^{(p)}(S)$ denote the value of $\mu_2(S)$ in ℓ_p^n . Statement (iii) of 6.1 is ready made to give us the value of $\mu_2^{(p)}(D_n)$:

<u>6.5.</u> $\mu_2^{(2)}(D_n) = \mu_2^{(\infty)}(D_n) = 2^{n/2}$, while $\mu_2^{(1)}(D_n) = \sqrt{n}\, 2^{n/2}$.

Proof. Consider ℓ_1^n (the other cases are similar). We have

$$[\, \mu_2^{(1)}(D_n)]^2 = \sup \{ \sum_{\varepsilon} \langle x,\varepsilon\rangle^2 : \|x\|_\infty \leqslant 1 \}$$

$$= 2^n \sup \{\|x\|_2^2 : \|x\|_\infty \leqslant 1 \}$$

$$= n2^n .$$

With this, we can calculate the summing norms in three more cases:

<u>6.6.</u> $\pi_2(I^{(n)}_{2,1}) = \pi_1(I^{(n)}_{2,1}) = \pi_2(I^{(n)}_{\infty,1}) = n$.

Proof. If J denotes any of these operators, then the expression $\Sigma\, e_i \otimes e_i$ shows that $\nu_1(J) \leqslant n$, so $\pi_p(J) \leqslant n$.

Now $\mu_2^{(2)}(D_n) = 2^{n/2}$, while

$$\sum_{\varepsilon\in D_n} \|\varepsilon\|_1^2 = n^2 2^n .$$

Hence $\pi_2(I^{(n)}_{2,1}) \geqslant n$. The other two quantities are clearly not less than $\pi_2(I^{(n)}_{1,2})$. (Alternatively, we have from 4.2 and 5.11 :

$$n = \mathrm{trace}(I_{1,1}) \leqslant \pi_2(I_{1,2})\, \pi_2(I_{2,1}) = \pi_2(I_{2,1}) \text{).}$$

<u>Exercise.</u> Give a direct proof that $\pi_2(I^{(n)}_{1,1}) = \sqrt{n}$ (of course, this is a special case of 5.13).

We summarize in tabular form what we now know about $I_{p,q}^{(n)}$ from the above results and those of sections 1 and 3 :

	ν_1			π_2			π_1		
	$q=\infty$	2	1	∞	2	1	∞	2	1
$p=\infty$	n	n	n	$\sqrt{n}$	$\sqrt{n}$	n	n	n	n
2	$\sqrt{n}$	n	n	$\sqrt{n}$	$\sqrt{n}$	n	$\sqrt{n}$		n
1	1	$\sqrt{n}$	n	1	1	$\sqrt{n}$	1		

The remaining cases for π_1 are not at all trivial; they will emerge from the results of sections 7 and 8. We leave it to the reader to summarize what we know so far about $\nu_\infty(I_{p,q}^{(n)})$.

Vector averaging and a property of Hilbert spaces

There are various ways to generalize 6.1. If we replace the scalars x_i, y_i by vectors and functionals, we obtain the following :

6.7 Proposition. Let $x_1, \dots, x_n$ be elements of a linear space X, and let $f_1, \dots, f_n$ be functionals on X. For $\varepsilon \in D_n$, write

$$y_\varepsilon = \sum_i \varepsilon_i x_i\ , \quad g_\varepsilon = \sum_i \varepsilon_i f_i\ .$$

Then

$$\frac{1}{2^n} \sum_{\varepsilon \in D_n} g_\varepsilon(y_\varepsilon) = \sum_i f_i(x_i)\ .$$

In particular, if X is an inner-product space, then

$$\frac{1}{2^n} \sum_{\varepsilon \in D_n} \|y_\varepsilon\|^2 = \sum_i \|x_i\|^2\ .$$

Proof. For a particular ε ,

$$g_\varepsilon(y_\varepsilon) = \sum_i f_i(x_i) + \sum_{i \neq j} \varepsilon_i \varepsilon_j\, f_i(x_j)\ .$$

Summation over ε gives the first statement, and the second statement follows at once.

6.8 Corollary. Given elements $x_1, \dots, x_n$ of an inner-product space, there exist $\delta, \varepsilon \in D_n$ such that

$$\| \Sigma\, \delta_i x_i \|^2 \leq \Sigma\, \|x_i\|^2 ,$$

$$\| \Sigma\, \varepsilon_i x_i \|^2 \geq \Sigma\, \|x_i\|^2 .$$

(This re-proves 2.12, and gives an opposite inequality as well).

6.9 Corollary. For any inner-product space H, $\pi_1^{(n)}(H) \leq \sqrt{n}$, with equality if $\dim H \geq n$.

Proof. Given n elements $x_1, \dots, x_n$, we have by 6.8,

$$\Sigma\, \|x_i\| \leq \sqrt{n}\, (\Sigma\, \|x_i\|^2)^{\frac{1}{2}} \leq \sqrt{n}\, \mu_1(x_1, \dots, x_n) .$$

Equality occurs for n orthonormal elements.

We shall see in section 8 that $\pi_1(\ell_2^n) > \sqrt{n}$, so $\pi_1(\ell_2^n) \neq \pi_1^{(n)}(\ell_2^n)$. This result has a nice application to Banach-Mazur distances :

6.10 Proposition. $d(\ell_\infty^n, \ell_2^n) = d(\ell_1^n, \ell_2^n) = \sqrt{n}$.

Proof. The identity operator shows that both quantities are not greater than $\sqrt{n}$. Equality for ℓ_∞^n follows from the fact that $\pi_1^{(n)}(\ell_\infty^n) = n$. The statement for ℓ_1^n follows by duality.

Of course, it is easy to deduce 6.10 from 6.8 without mentioning $\pi_1^{(n)}$. For example, let $S : \ell_1^n \to \ell_2^n$ be an isomorphism with $\|S\| = 1$. Then there exists $\delta \in D_n$ with $\|\Sigma\, \delta_i(Se_i)\|_2 \leq \sqrt{n}$. Since $\|\Sigma\, \delta_i e_i\|_1 = n$, this shows that $\|S^{-1}\| \geq \sqrt{n}$.

Exercise. With the notation of 6.7, show that

$$\frac{1}{2^n} \sum_{\varepsilon} g_\varepsilon \otimes y_\varepsilon = \sum_{i} f_i \otimes x_i$$

and

$$\mu_2\{y_\varepsilon : \varepsilon \in D_n\} = 2^{n/2} \mu_2(x_1, \dots, x_n) .$$

(These statements generalize 6.1 (i) and 6.5 respectively).

Exercise. For any operator T on ℓ_1^n, show that

$$\pi_2(T) \leq \mu_2(Te_1, \dots, Te_n).$$

Type 2 and cotype 2 constants

Given a finite sequence $(x_1, \dots, x_n)$ of elements of a normed linear space, write

$$\rho_2(x_1, \dots, x_n) = \left[\frac{1}{2^n} \sum_{\varepsilon \in D_n} \|y_\varepsilon\|^2\right]^{\frac{1}{2}} ,$$

where $y_\varepsilon = \Sigma\, \varepsilon_i x_i$. (The notation ρ_2, like μ_p, is not standard).

Proposition 6.7 says that for elements of an inner product space,

$$\rho_2(x_1, \dots, x_n) = (\Sigma \|x_i\|^2)^{\frac{1}{2}} .$$

Motivated by this, we define the type 2 constant τ_2 and the cotype 2 constant κ_2 of a normed linear space X as follows:

$$\tau_2(X) = \sup \{\rho_2(x_1, \dots, x_n) : \Sigma \|x_i\|^2 \leqslant 1\} ,$$

$$\kappa_2(X) = \sup \{(\Sigma \|x_i\|^2)^{\frac{1}{2}} : \rho_2(x_1, \dots, x_n) \leqslant 1\} .$$

The space X is said to be of type or cotype 2 if the corresponding constant is finite. Clearly we have $\tau_2(H) = \kappa_2(H) = 1$ for any inner-product space H. At this stage, we restrict our account to the really elementary facts related to these concepts. The connection with summing and nuclear norms will unfold in later sections. First, some immediate remarks on the definitions

(1) If T is an operator, then $\rho_2(Tx_1, \dots, Tx_n) \leqslant \|T\|\, \rho_2(x_1, \dots, x_n)$.

(2) If X is a subspace of X, then $\tau_2(X_1) \leqslant \tau_2(X)$, $\kappa_2(X_1) \leqslant \kappa_2(X)$. Since the definitions involve only finitely many elements, we have

$$\tau_2(X) = \sup \{\tau_2(X_1) : X_1 \text{ a finite-dimensional subspace X}\} ,$$

and similarly for κ_2. Further, if Y is finitely represented in X (for the meaning of this, see section 0), then $\tau_2(Y) \leqslant \tau_2(X)$, etc.

(3) $\rho_2(x_1, \dots, x_n) \leqslant \mu_1(x_1, \dots, x_n)$, since $\|y_\varepsilon\| \leqslant \mu_1(x_1, \dots, x_n)$ for each ε. Hence $\pi_{2,1}(I_X) \leqslant \kappa_2(X)$. In words, cotype 2 implies the Orlicz property.

(4) Clearly, the definitions of τ_2, κ_2 can be applied to linear operators : we simply replace x_i by Tx_i on the left-hand side. Then $\tau_2(X)$ equates to $\tau_2(I_X)$. In particular, κ_2 can be thought of as a modification of π_2, in which ρ_2 replaces μ_2. The statement in (3) becomes $\pi_{2,1}(T) \leqslant \kappa_2(T)$.

(5) An alternative expression for $\rho_2(x_1, \dots, x_n)^2$, which the reader will often encounter in the literature, is

$$\int_0^1 \|\sum_1 r_i(t)\, x_i\|^2\, dt ,$$

where $r_1, \ldots, r_n$ are the first in Rademacher functions on [0,1]. This is the same, since for each $\varepsilon \in D_n$, the set $\{t : r_i(t) = \varepsilon_i$ for each $i\}$ is an interval of length $1/2^n$.

(6) The definitions only use real scalars.

<u>6.11.</u> $\rho_2(x_1, \ldots, x_n) \geq \mu_2(x_1, \ldots, x_n)$. Hence $\kappa_2(T) \leq \pi_2(T)$ for any operator T, and if $\dim X = n$, then $\kappa_2(X) \leq \sqrt{n}$.

Proof. Take $f \in U_{X^*}$. We have $f(y_\varepsilon) = \Sigma \varepsilon_i f(x_i)$, and hence by 6.1(iii),

$$\sum_1 f(x_i)^2 = \frac{1}{2^n} \sum_\varepsilon f(y_\varepsilon)^2 \leq \frac{1}{2^n} \sum_\varepsilon \|y_\varepsilon\|^2 = \rho_2(x_1, \ldots, x_n)^2 .$$

<u>6.12.</u> For isomorphic spaces X,Y, we have

$$\tau_2(Y) \leq d(X,Y)\, \tau_2(X), \qquad \kappa_2(Y) \leq d(X,Y)\, \kappa_2(X) .$$

For τ_2 (but not κ_2) we have the following stronger statement : if there is an M-open operator τ of X onto Y with $\|\tau\| = 1$, then $\tau_2(Y) \leq M\, \tau_2(X)$. Hence if E is a subspace of X, then $\tau_2(E^*) \leq \tau_2(X^*)$.

Proof. We just prove the "stronger statement" for τ_2. Take elements y_i of Y with $\Sigma \|y_i\|^2 \leq 1$. There are elements x_i of X with $Tx_i = y_i$ and $\|x_i\| \leq M\|y_i\|$. Then $\rho_2(x_1, \ldots, x_n) \leq M\tau_2(X)$. Since $\|T\| \leq 1$, it follows that $\rho_2(y_1, \ldots, y_n) \leq M\tau_2(X)$.

The reader should reflect on why a similar argument cannot be used to prove the "stronger statement" for κ_2. Examples below will show that no such statement is true.

Clearly we have from 6.12 and 5.6 that for any n-dimensional space X, $\tau_2(X) \leq \sqrt{n}$ and (again) $\kappa_2(X) \leq \sqrt{n}$.

<u>6.13 Example</u> $\kappa_2(\ell_\infty^n) = \tau_2(\ell_1^n) = \sqrt{n}$.

$\tau_2(\ell_\infty^n) \to \infty$ as $n \to \infty$.

Proof. In ℓ_∞^n, $\|\Sigma\, \varepsilon_i e_i\| = 1$ for all ε, so $\rho_2(e_1, \ldots, e_n) = 1$, while $\Sigma \|e_i\|^2 = n$.

In ℓ_1^n, we have $\rho_1(e_1, \dots, e_n) = n$, while $(\Sigma \|e_i\|^2)^{1/2} = \sqrt{n}$.

The final statement follows from the fact that ℓ_1^n embeds isometrically into ℓ_∞^N for a suitable N. This is achieved (in the real case) by associating with $x \in \ell_1^n$ the function $\hat{x}$ on D_n defined by : $\hat{x}(\varepsilon) = \langle x, \varepsilon\rangle$. Hence $N = 2^n$, and we have shown that $\tau_2(\ell_\infty^N) \geqslant n$.

The most interesting result of this sort on elementary spaces is that ℓ_1 is of cotype 2. This will be proved in section 7.

6.14. For any normed linear space X,

$$\kappa_2(X^*) \leqslant \tau_2(X), \qquad \kappa_2(X) \leqslant \tau_2(X^*) .$$

Proof. The second statement follows from the first, since $\kappa_2(X) \leqslant \kappa_2(X^{**})$.

Let $\tau_2(X) = \alpha$. Take $f_1, \dots, f_n \in X^*$ and $\delta > 0$. There are elements x_i of X with $\|x_i\| = \|f_i\|$ and $f_i(x_i) \geqslant (1+\delta)^{-1}\|f_i\|^2$. Write $y_\varepsilon = \Sigma\, \varepsilon_i x_i$, $g_\varepsilon = \Sigma\, \varepsilon_i f_i$. Then

$$\begin{aligned}\sum_i \|f_i\|^2 &\leqslant (1+\delta) \sum_i f_i(x_i) \\ &= \frac{1+\delta}{2^n} \sum_\varepsilon g_\varepsilon(y_\varepsilon) \qquad\qquad \text{by 6.7} \\ &\leqslant \frac{1+\delta}{2^n} \Big(\sum_\varepsilon \|g_\varepsilon\|^2\Big)^{1/2} \Big(\sum_\varepsilon \|y_\varepsilon\|^2\Big)^{1/2} \\ &\leqslant (1+\delta)\, 2^{-n/2} \Big(\sum_\varepsilon \|g_\varepsilon\|^2\Big)^{1/2} \alpha \Big(\sum_i \|x_i\|^2\Big)^{1/2}\end{aligned}$$

by the definition of α. Since $\|x_i\| = \|f_i\|$, this gives

$$(\Sigma \|f_i\|^2)^{1/2} \leqslant (1+\delta)\, \alpha\, \rho_2(f_1, \dots, f_n) .$$

Hence $\kappa_2(X^*) \leqslant (1+\delta)\alpha$.

In exactly the same way, one has $\kappa_2(T^*) \leqslant \tau_2(T)$ for an operator T. The above remark about ℓ_1 shows that τ_2 and κ_2 cannot be interchanged.

For general p, "type p" and "cotype p" constants are defined in the same way, replacing $(\Sigma \|x_i\|^2)^{1/2}$ by $(\Sigma \|x_i\|^p)^{1/p}$, but leaving ρ_2 unchanged (actually, ρ_p is "equivalent" to ρ_2). Clearly, even $\mathbb{R}^1$ is only of type p for $p \leqslant 2$ and cotype p for $p \geqslant 2$. The study of these concepts was initiated by Maurey & Pisier (see especially (1976)).

Averaging operators : a general result on trace duality

We now describe one way in which the averaging notion can be applied to operators. Roughly speaking, "averaging" any operator on K^n reduces it to a multiple of the identity. This leads to a general trace duality result on $I_{p,q}^{(n)}$. The idea is due to Garling & Gordon (1971). As before, K denotes $\mathbb{R}$ or $\mathbb{C}$.

6.15. Let S be be any operator in $L(K^n)$, and let $Se_j = \sum_i \alpha_{ij} e_i$ for each j. For ε in D_n, let U_ε be the operator $\operatorname{diag}(\varepsilon_1, \ldots, \varepsilon_n)$. Let

$$S_0 = \frac{1}{2^n} \sum_{\varepsilon \in D_n} U_\varepsilon S U_\varepsilon .$$

Then $S_0 = \operatorname{diag}(\alpha_{11}, \ldots, \alpha_{nn})$.

Proof. For each j, we have $SU_\varepsilon e_j = \varepsilon_j S e_j$, and hence

$$U_\varepsilon S U_\varepsilon e_j = \varepsilon_j U_\varepsilon S e_j = \varepsilon_j \sum_i \alpha_{ij} \varepsilon_i e_i$$

$$= \alpha_{jj} e_j + \sum_{i \neq j} \varepsilon_i \varepsilon_j \alpha_{ij} e_i .$$

Summation over $\varepsilon \in D_n$ gives the statement.

6.16. Let S, S_0 be as in 6.15, and define $H \in L(K^n)$ by : $He_i = e_{i+1}$ for $1 \leqslant i \leqslant n-1$, $He_n = e_1$. Then

$$\sum_{r=1}^{n} H^{-r} S_0 H^r = (\text{trace } S)\, I^{(n)} .$$

Proof. For any permutation σ of $\{1, \ldots, n\}$, let H_σ be the operator defined by : $H_\sigma e_i = e_{\sigma(i)}$. Then $H_\sigma^{-1} S_0 H_\sigma e_i = \alpha_{\sigma(i),\sigma(i)} e_i$. Now let σ be the particular permutation given by $\sigma(i) = i+1$ for $i < n$, $\sigma(n) = 1$. The elements $\sigma(i), \sigma^2(i), \ldots, \sigma^n(i)$ are $1, \ldots, n$ in some order. The statement follows.

6.17 Proposition. Let α be any operator ideal norm, α^* its dual under finite-dimensional trace duality. Then, for any p,q,n,

$$\alpha(I_{p,q}^{(n)})\, \alpha^*(I_{q,p}^{(n)}) = n .$$

Proof. The quantity under consideration is of course not less than n. There is an element S of $L(\ell_p^n, \ell_q^n)$ such that $\alpha(S) = 1$ and trace S =

$\alpha^*(I^{(n)}_{q,p})$. In both ℓ^n_p and ℓ^n_q, we have (with the above notation) $\|U_\varepsilon\| = \|H\| = 1$. Hence $\alpha(U_\varepsilon S U_\varepsilon) \leq 1$ (as an operator from ℓ^n_p to ℓ^n_q), and $\alpha(S_0) \leq 1$. So by 6.16,

$$(\text{trace } S)\ \alpha(I^{(n)}_{p,q}) \leq \sum_r \alpha(H^{-r} S_0 H^r) \leq n\ \alpha(S_0) \leq n .$$

This proves the statement.

<u>6.18 Corollary.</u> For each p and n, $\lambda(\ell^n_p)\,\pi_1(\ell^n_p) = n$.

<u>6.19 Corollary.</u> Suppose that $d(X, \ell^n_p) = C$ for some p. If α, α^* are as in 6.17, then $\alpha(I_X)\alpha^*(I_X) \leq C^2 n$.

Proof. We have $\alpha(I_X) \leq C\alpha(I^{(n)}_{p,p})$, and similarly for α^*.

<u>6.20 Example.</u> Let $X = \ell^n_\infty \times \ell^n_1$, with norm defined by : $\|(x,y)\| = \max(\|x\|, \|y\|)$. Then $\dim X = 2n$ and $\pi_1(X) \geq \pi_1(\ell^n_\infty) = n$. As we will see in section 7, $\lambda(\ell^n_1) \geq \sqrt{n/2}$, and hence we have $\lambda(X)\pi_1(X) \geq n\sqrt{n/2}$.

7. MORE AVERAGING : KHINCHIN'S INEQUALITY AND RELATED RESULTS

Khinchin's inequality

Section 6 was concerned with statements derived from the easy "averaging" result 6.1, or variants of it (all equalities). We now prove an inequality that is distinctly less easy; it was discovered by Khinchin as long ago as 1923. (The spelling Khinchin is the correct romanization according to the system normally used in English. However, the French version Khintchine is often encountered). The inequality deals with the average of $|\langle \varepsilon, x\rangle|$ instead of $\langle \varepsilon, x\rangle^2$. As we shall see, it is of fundamental importance in the study of π_1, projection constants and cotype 2 constants.

Let $x = (x_1, \dots, x_n)$ be an element of $\mathbb{R}^n$ (in this section, we write x_i rather than $x(i)$). Write

$$\rho_1(x) = \frac{1}{2^n} \sum_{\varepsilon \in D_n} |\langle \varepsilon, x\rangle| .$$

First, note that $\rho_1(x) \leq \|x\|_2$. This follows from 6.1(iii) and the elementary fact that for positive numbers $a_1, \dots, a_N$,

$$\frac{1}{N}(a_1+\dots+a_N) \leq \left[\frac{1}{N}(a_1^2+\dots+a_N^2)\right]^{\frac{1}{2}}$$

(proof : $\sum_j (a_j-c)^2 \geq 0$, with $c = \frac{1}{N}(a_1+\dots+a_N)$) .

Khinchin's inequality is the remarkable statement that, conversely, $\rho_1(x) \geq C\|x\|_2$ for a certain C independent of n and x. Before proving it, we note that a very simple example is enough to show that $C \leq 1/\sqrt{2}$.

<u>7.1 Example</u> Let $x = (1,1) \in \mathbb{R}^2$. Then $\rho_1(x) = \frac{1}{2}(2+0) = 1$, while $\|x\|_2 = \sqrt{2}$.

We will give two proofs of the inequality. One of them depends on the following lemma (which is used again in section 11).

7.2 Lemma. $\frac{1}{2^n} \sum_{\varepsilon \in D_n} \langle \varepsilon, x \rangle^4 \leqslant 3\|x\|_2^4$.

Proof. We have

$$\|x\|_2^4 = (\sum_i x_i^2)^2 = \sum_i x_i^4 + 2 \sum_{i<j} x_i^2 x_j^2 .$$

In the sum $\sum_\varepsilon \langle \varepsilon, x\rangle^4$, summation over ε gives 0 for the terms involving $\varepsilon_i \varepsilon_j^3$ $(=\varepsilon_i\varepsilon_j)$, $\varepsilon_i\varepsilon_j\varepsilon_k^2$ $(=\varepsilon_i\varepsilon_j)$ and $\varepsilon_i\varepsilon_j\varepsilon_k\varepsilon_\ell$ (i,j,k,ℓ distinct). The remaining terms in $\langle \varepsilon, x\rangle^4$ are

$$\sum_i x_i^4 + 6 \sum_{i<j} x_i^2 x_j^2$$

(note that $\binom{4}{2} = 6$). This expression is therefore the average of $\langle \varepsilon, x\rangle^4$, and the statement follows.

7.3 Theorem. There exists $C \geqslant 1/\sqrt{3}$ such that for any n and any $x \in \mathbb{R}^n$,

$$C \|x\|_2 \leqslant \rho_1(x) \leqslant \|x\|_2 ,$$

where ρ_1 is defined as above.

Proof 1. For all real t, we have $|t| \geqslant \frac{3}{2} t^2 - \frac{1}{2} t^4$, since the greatest value of $t - \frac{1}{3}t^3$ for $t > 0$ occurs at $t = 1$, and is $\frac{2}{3}$. Hence

$$\rho_1(x) \geqslant \frac{3}{2} \frac{1}{2^n} \sum_\varepsilon \left[\langle \varepsilon, x\rangle^2 - \frac{1}{3} \langle \varepsilon, x\rangle^4 \right]$$

$$\geqslant \frac{3}{2} (\|x\|^2 - \|x\|^4)$$

by 6.1 and 7.2 (we write $\| \ \|$ for $\| \ \|_2$). Hence

$$\frac{\rho_1(x)}{\|x\|} \geqslant \frac{3}{2} (\|x\| - \|x\|^3) .$$

Since $\rho_1(x)/\|x\|$ is unchanged if x is replaced by λx, we have

$$\frac{\rho_1(x)}{\|x\|} \geqslant \frac{3}{2} (u - u^3)$$

for all $u > 0$. The greatest value of $u - u^3$ occurs when $3u^2 = 1$, giving

$$\frac{\rho_1(x)}{\|x\|} \geqslant \frac{3}{2} \frac{1}{\sqrt{3}} \frac{2}{3} = \frac{1}{\sqrt{3}} .$$

Proof 2. This uses complex numbers (although the conclusion, again, is only for <u>real</u> x_j). Write

$$g(\varepsilon) = (1 + i\varepsilon_1 x_1) \dots (1 + i\varepsilon_n x_n) .$$

Then

$$|g(\varepsilon)|^2 = (1 + x_1^2) \dots (1 + x_n^2)$$

$$\leq e^{x_1^2 + \dots + x_n^2} ,$$

so $|g(\varepsilon)| \leq e^{\frac{1}{2}\|x\|^2}$.

Consider the sum $\sum_{\varepsilon} \langle\varepsilon,x\rangle g(\varepsilon)$. It is clear that the only term that does not give 0 when summed over all ε is $i(x_1^2 + \dots + x_n^2)$. Hence

$$\frac{1}{2^n} \sum_{\varepsilon} \langle\varepsilon,x\rangle g(\varepsilon) = i\|x\|^2 ,$$

and therefore

$$\|x\|^2 \leq \frac{1}{2^n} \sum_{\varepsilon} |\langle\varepsilon,x\rangle \, g(\varepsilon)|$$

$$\leq e^{\frac{1}{2}\|x\|^2} \rho_1(x) .$$

So if $\|x\| = 1$, then $\rho_1(x) \geq 1/\sqrt{e}$ (note : $te^{-\frac{1}{2}t^2}$ is greatest when $t = 1$). This proves the theorem, this time with constant $1/\sqrt{e}$.

<u>Notes</u> (1) One can check that 7.2 and Proof 1 adapt for complex scalars. However, we will deduce the complex case as part of a more general extension to Hilbert spaces.

(2) For many years, the best constant known for the inequality was the $1/\sqrt{3}$ obtained in proof 1. It was eventually shown by Szarek (1976) that the best constant is in fact $1/\sqrt{2}$, so that the simple example in 7.1 is already the extreme case. Known proofs of this fact are surprisingly hard, and will not be reproduced here (see, for example, Haagerup (1982)). However, in stating further results that depend on Khinchin's inequality, we shall allow ourselves to assume the inequality with constant $1/\sqrt{2}$.

(3) There are in fact "Khinchin inequalities" for each $p \geq 1$. Write

$$\rho_p(x) = \left(\frac{1}{2^n} \sum_{\varepsilon} |\langle\varepsilon,x\rangle|^p \right)^{1/p} .$$

Then there are constants A_p, B_p such that

$$A_p \|x\|_2 \leqslant \rho_p(x) \leqslant B_p \|x\|_2$$

for all $x \in \mathbb{R}^n$ (it is trivial that $A_p = 1$ for $p \geqslant 2$ and $B_p = 1$ for $p \leqslant 2$). Note that 7.2 is the case $p = 4$; the proof is an extension of this method, and shows that $B_p \leqslant C\sqrt{p}$ for a constant C. See for example [CBS I], 2b. Using this, one can extend some of the applications below to p other than 1,2.

Exercise. Fill in the details of the following proof of Khinchin's inequality (with the right constant $1/\sqrt{2}$) for $n \leqslant 4$. Assume $x_1 \geqslant \dots \geqslant x_4 \geqslant 0$, $\|x\|_2 = 1$. Squaring shows that $3x_1 + x_2 + x_3 + x_4 \geqslant 2\sqrt{2}$. Let A_k be the following eight elements of D_4 : those ε with at most one -1, and those with $\varepsilon_1 = 1$ and two -1's. These add up to (6,2,2,2), so

$$\frac{1}{8} \sum_{\varepsilon \in A_4} |\langle \varepsilon, x\rangle| \geqslant \frac{1}{8} \sum_{\varepsilon \in A_4} \langle \varepsilon, x\rangle \geqslant \frac{1}{4} (3x_1 + x_2 + x_3 + x_4) \geqslant \frac{1}{\sqrt{2}} .$$

Exercise. Show that if $0 \leqslant x \leqslant y$, then $\rho_1(x) \leqslant \rho_1(y)$. (Consider the case where only one coordinate is different).

Hilbert space version and cotype 2 constant of ℓ_1

Let $x_1, \dots ,x_n$ be elements of a normed linear space X. As before, write $y_\varepsilon = \sum \varepsilon_i x_i$ and

$$\rho_1(x_1, \dots ,x_n) = \frac{1}{2^n} \sum_{\varepsilon \in D_n} \|y_\varepsilon\| .$$

Then, as for scalars, $\rho_1(x_1, \dots ,x_n) \leqslant \rho_2(x_1, \dots ,x_n)$.

Now consider $\mathbb{R}^k$ with the natural ordering and any lattice norm (see section 0), in particular any $\| \;\|_p$. Given elements x_i, there is an obvious "pointwise" definition of the element

$$z = (\sum_i x_i^2)^{1/2} .$$

By applying Khinchin's inequality to each coordinate, we have

$$0 \leqslant z \leqslant \frac{\sqrt{2}}{2^n} \sum_\varepsilon |y_\varepsilon| ,$$

and hence

$$\|z\| \leqslant \frac{\sqrt{2}}{2^n} \sum \|y_\varepsilon\| = \sqrt{2}\, \rho_1(x_1, \dots ,x_n). \tag{1}$$

In ℓ_2^k we have, clearly, $\|z\|^2 = \sum_i \|x_i\|^2$. Hence we have proved the following extension of Khinchin's inequality to inner-product spaces :

7.4. Let $x_1, \ldots, x_n$ be elements of a (real) inner-product space. Then

$$\frac{1}{\sqrt{2}} (\sum_i \|x_i\|^2)^{1/2} \leqslant \rho_1(x_1, \ldots, x_n) \leqslant (\sum_i \|x_i\|^2)^{1/2} .$$

In particular, Khinchin's inequality holds for complex numbers.

Proof. We may assume that the space is ℓ_2^n. The above remarks prove the left-hand inequality, and the right-hand one follows from $\rho_2(x_1, \ldots, x_n) = (\Sigma \|x_i\|^2)^{1/2}$ (6.7). The statement for complex numbers follows, since $\mathbb{C} = \ell_2^2$.

The statement holds in fact for complex inner-product spaces too, since a real inner product, inducing the same norm, is defined on such a space by $\langle x,y\rangle_R = \mathrm{Re}\langle x,y\rangle$.

In order to apply (1) to ℓ_1, we need information about $\|z\|$. This is provided by the following simple lemma : the property of ℓ_1 it describes is known as "2-concavity".

7.5 Lemma. Let $x_1, \ldots, x_n$ be elements of (real) ℓ_1^k or ℓ_1, and let $z = (\sum_i x_i^2)^{1/2}$. Then $\|z\|^2 \geqslant \Sigma \|x_i\|^2$.

Proof. Let $\|x_i\| = \lambda_i$. By Schwarz's inequality,

$$\sum_i \lambda_i |x_i| \leqslant (\sum_i \lambda_i^2)^{1/2} z .$$

Now

$$\|\sum_i \lambda_i |x_i| \| = \sum_i \lambda_i \|x_i\| = \sum_i \lambda_i^2 ,$$

so

$$(\sum_i \lambda_i^2)^{1/2} \leqslant \|z\| .$$

Notes (1) Clearly, this applies also to complex ℓ_1, with $z = (\Sigma |x_i|^2)^{1/2}$ (note that $|x_i|$ and z are still elements of real ℓ_1 !).

(2) Similar reasoning shows that $\phi(z)^2 \geq \Sigma\, \phi(x_i)^2$ for any positive functional on $\mathbb{R}^k$.

7.6 Proposition. Let X be ℓ_1^k or ℓ_1 (real or complex; $k \geq 2$). Then $\pi_{2,1}(X) \leq \kappa_2(X) = \sqrt{2}$. In the real case, $\pi_{2,1}(X) = \sqrt{2}$.

Proof. By (1) and 7.5, we have

$$(\Sigma \|x_i\|^2)^{\frac{1}{2}} \leq \|z\| \leq \sqrt{2}\, \rho_1(x_1, \ldots, x_n)$$

so that $\kappa_2(X) \leq \sqrt{2}$ (note that our statement is slightly stronger, in that ρ_1 appears instead of ρ_2). Of course, $\pi_{2,1}(X) \leq \kappa_2(X)$.

To prove equality where stated, let $x_1 = (1,1)$, $x_2 = (1,-1)$. Then $\|x_1 + x_2\| = \|x_1 - x_2\| = 2$, so $\rho_2(x_1, x_2) = 2$, and $\mu_1(x_1, x_2) = 2$ in the real case. But $(\|x_1\|^2 + \|x_2\|^2)^{\frac{1}{2}} = 2\sqrt{2}$.

Of course, the same applies to any subspace of ℓ_1, and to any space (like $L_1(\mu)$ or $C(S)^*$) that is finitely represented in ℓ_1 (for $L_1(\mu)$, the above proof can be applied directly). As mentioned earlier, the statement for $\pi_{2,1}$ was proved by Orlicz (1933), in what can be regarded as the article that originated the whole theory of summing operators.

7.7 Corollary. If E is any n-dimensional subspace of ℓ_1, then $d(E, \ell_\infty^n) \geq \sqrt{n/2}$.

Proof. $\kappa_2(E) \leq \sqrt{2}$, while $\kappa_2(\ell_\infty^n) = \sqrt{n}$.

Notes. (1) Since $z = \frac{1}{2^n} (\Sigma\, y_\varepsilon^2)^{\frac{1}{2}}$, we have from 7.5 that $\|z\| \geq \rho_2(x_1, \ldots, x_n)$, a reverse inequality for $\|z\|$.

(2) In the same way, one can show that ℓ_p is of cotype 2 for $p \leq 2$ and type 2 for $p \geq 2$ (in particular, Hilbert spaces are not the only spaces of type 2).

7.8 Example. As far as the author knows, the exact value of $\pi_{2,1}$ for complex ℓ_1 is not known. We give a bare sketch of the steps required to prove that $\pi_{2,1}(\ell_1^2) = 1$ and $\pi_{2,1}(\ell_1^3) \geq 3\sqrt{2}/4$ in the complex case; the reader may care to fill in the details.

To show $\pi_{2,1}(X) = 1$, it is sufficient to show that given elements x,y, there exists α with $|\alpha| = 1$ and $\|x + \alpha y\|^2 \geq \|x\|^2 + \|y\|^2$ (1). Let

$x = (x_1, \rho x_2)$, $y = (y_1, \rho\sigma y_2)$ with x_j, y_j real, positive and $|\rho| = |\sigma| = 1$. Let $\sigma = e^{2i\theta}$ (with $-\pi/2 \leqslant \theta \leqslant \pi/2$), $\alpha = e^{-i\theta}$. Then (1) holds.

For ℓ_1^3, let $x = (1,1,1)$, $y = (\beta, -1, \bar{\beta})$, where $\beta = e^{i\pi/3}$. The statement follows if we can show $\mu_1(x,y) = 4$. To prove this, we use the fact that $|e^{i\theta} - e^{i\phi}| = 2\,|\sin \tfrac{1}{2}(\theta - \phi)|$ to prove that if $\alpha = e^{i\theta}$, where $|\theta| \leqslant \pi/3$, then $|\alpha - \beta| + |\alpha - \bar{\beta}| \leqslant 2$.

Khinchin's inequality also shows that (apart from the factor $\sqrt{2}$) ℓ_2 is finitely represented in ℓ_1. More precisely.

7.9. Let $N = 2^n$. Then there is a subspace E_n of ℓ_1^N such that $d(E_n, \ell_2^n) \leqslant \sqrt{2}$.

Proof. Let Y be the space of all functions y on the set D_n, with norm defined by : $\|y\| = \frac{1}{2^n} \sum_{\varepsilon} |y(\varepsilon)|$. Clearly, Y is isometric to ℓ_1^N. For x in ℓ_2^n, define $Jx \in Y$ by : $(Jx)(\varepsilon) = \langle x, \varepsilon \rangle$. Khinchin's inequality gives

$$\frac{1}{\sqrt{2}} \|x\|_2 \leqslant \|Jx\| \leqslant \|x\|_2 .$$

Applications to 1-summing norms and projection constants

Khinchin's inequality enables us to provide estimates (either exact or in the form of inequalities) for $\pi_1(I_{p,q}^{(n)})$ in the cases left unsolved in section 6. In each case, the answer is extreme in the sense of being close to the smallest possible, and we conclude that the projection constants of ℓ_1^n and ℓ_2^n are close to being the largest possible (i.e. $\sqrt{n}$) for n-dimensional spaces.

7.10 Proposition. $\pi_1(I_{1,2}^{(n)}) \leqslant \sqrt{2}$ for all $n \geqslant 2$; $\pi_1(I_{1,2}) \leqslant \sqrt{2}$. Equality holds in the real case.

Proof. By 7.3 (also valid in the complex case),

$$\|x\|_2 \leqslant \frac{\sqrt{2}}{2^n} \sum_{\varepsilon} |\langle \varepsilon, x \rangle| .$$

As a functional on ℓ_1^n, $\|\varepsilon\| = 1$. The statements follow, by 3.2.

To prove equality for the real case with n=2, let $x_1 = (1,1)$, $x_2 = (1,-1)$. Then $\mu_1(x_1, x_2) = 2$ in ℓ_1^2 (as in 7.6), while $\|x_1\|_2 + \|x_2\|_2 = 2\sqrt{2}$.

7.11 Proposition. $\pi_1(\ell_2^n) \leqslant \sqrt{2n}$, $\lambda(\ell_2^n) \geqslant \sqrt{n/2}$. The space ℓ_2 is not injective.

Proof. The statement for π_1 is proved as in 7.10; we now have $\|\mathcal{E}\| = \sqrt{n}$.

If $\dim X = n$, then $\lambda(X)\pi_1(X) \geqslant n$, by 3.8. Hence $\lambda(\ell_2^n) \geqslant \sqrt{n/2}$, and ℓ_2 is not injective.

7.12 Proposition. $\pi_1(\ell_1^n) \leqslant \sqrt{2n}$, $\lambda(\ell_1^n) \geqslant \sqrt{n/2}$. The space ℓ_1 is not injective.

Proof. Again as in 7.10, using the fact that $\|x\|_1 \leqslant \sqrt{n}\ \|x\|_2$. The other statements follow, as in 7.11.

7.13 Example. For the complex case, example 7.8 adapts easily to show that $\pi_1(I_{1,2}^{(2)}) \geqslant 3\sqrt{2}/4$ (though this is certainly not the exact value !). With the notation of 7.8, let $a_1 = (1, \beta)$, $a_2 = (1,-1)$, $a_3 = (1, \bar{\beta})$. In the same way (or by 2.11), we have $\mu_1(a_1,a_2,a_3) = 4$ in ℓ_1^2 , while $\Sigma\ \|a_i\|_2 = 3\sqrt{2}$.

Exact evaluation of constants for ℓ_1^n

For interest, we now show how one can determine π_1 and λ exactly for (real) ℓ_1^n. Of course, the reader who is content with the estimate in 7.12 can leave this out. The evaluation of $\lambda(\ell_1^n)$ was first achieved by B. Grünbaum (1960); the method presented here is essentially that of Y. Gordon (1969). It depends on a direct comparison of $\rho_1(x)$ and $\|x\|_1$ (not $\|x\|_2$), independent of Khinchin's inequality.

It is clear that ρ_1 is a norm and that $\rho_1(|x|) = \rho_1(x)$. Write $C_n = \rho_1(e)$, where $e = (1, \dots ,1) \in \mathbb{R}^n$.

From the expression $I^{(n)} = 2^{-n} \sum_{\mathcal{E}} \mathcal{E} \otimes \mathcal{E}$ (6.1), we have

$$\lambda(\ell_1^n) = \nu_\infty(I_{1,1}^{(n)}) \leqslant 2^{-n}\mu_1^{(1)}(D_n) .$$

7.14 Lemma. $\lambda(\ell_1^n) \leqslant C_n$.

Proof. D_n is a norming subset of the unit ball of ℓ_∞^n (regarded as the dual of ℓ_1^n), so

$$\mu_1^{(1)}(D_n) = \sup \{\sum_{\mathcal{E}\in D_n} |\langle\mathcal{E},\delta\rangle| : \delta \in D_n\}$$

$$= 2^n \sup \{\rho_1(\delta) : \delta \in D_n\} .$$

But $\rho_1(\delta) = \rho_1(e) = C_n$ for each $\delta \in D_n$. Hence $\mu_1^{(1)}(D_n) = 2^n C_n$.

7.15 Lemma. $\rho_1(x) \geq \frac{1}{n} C_n \|x\|_1$ for $x \in \mathbb{R}^n$.

Proof. We assume $x \geq 0$, since $\rho_1(|x|) = \rho_1(x)$. Let $Hx = (x_2, \ldots, x_n, x_1)$, where $x = (x_1, \ldots, x_n)$. Clearly, $\rho_1(Hx) = \rho_1(x)$. Now

$$x + Hx + \ldots + H^{n-1}x = \|x\|_1 e .$$

Since ρ_1 is sub-additive, it follows that

$$n\rho_1(x) \geq \|x\|_1 \rho_1(e) = \|x\|_1 C_n$$

Note. This shows that the least value of $\rho_1(x)$ when $\|x\|_1 = 1$ occurs at e/n. Of course, the greatest value is 1, attained at each e_j.

7.16 $\lambda(\ell_1^n) = C_n$, $\pi_1(\ell_1^n) = \dfrac{n}{C_n}$.

Proof. The previous lemma says that

$$\|x\|_1 \leq \frac{n}{C_n} \frac{1}{2^n} \sum_{\varepsilon} |\langle \varepsilon, x \rangle| ,$$

so by 3.2, $\pi_1(\ell_1^n) \leq \dfrac{n}{C_n}$. Since $\lambda(\ell_1^n)\, \pi_1(\ell_1^n) \geq n$ (by 3.8), equality follows. (Note that 6.18 is not needed).

Remark. D_n itself is a finite set of which $\pi_1(\ell_1^n)$ is attained, since $\mu_1(D_n) = 2^n C_n$ (see 7.14), while $\Sigma\, \{\|\varepsilon\|_1 : \varepsilon \in D_n\} = n2^n$.

We now tackle the essentially combinatorial problem of evaluating C_n.

7.17 Proposition. We have

$$C_{2n-1} = C_{2n} = \frac{1.3.\ldots.(2n-1)}{2.4.\ldots.(2n-2)}$$

and
$$\pi_1(\ell_1^{2n}) = \pi_1(\ell_1^{2n+1}) = \frac{2.4 \ldots (2n)}{1.3 \ldots (2n-1)} = (2n+1) \int_0^{\pi/2} \cos^{2n+1}\theta \, d\theta .$$

Let $K_n = C_n/\sqrt{n}$. Then $K_{2n} \uparrow \sqrt{2/\pi}$ and $K_{2n-1} \downarrow \sqrt{2/\pi}$ as $n \to \infty$.

Proof. Write

$$C_n' = 2^n C_n = \sum_{\varepsilon} |\langle \varepsilon, e\rangle| = \sum_{\varepsilon} |\sum_i \varepsilon_i| .$$

Firstly, we show that $C_{2n}' = 2C_{2n-1}'$. Consider an element ε of D_{2n-1} with $\sum^{2n-1} \varepsilon_i \geq 0$. This sum is then at least 1, and the choices $\varepsilon_{2n} = 1, -1$ give two extensions of ε to elements of D_{2n}, which we denote by ε^+, ε^-. Both have non-negative sum, and

$$\sum_1^{2n} \varepsilon_i^+ = \sum_1^{2n} \varepsilon_i^- = 2 \sum_1^{2n-1} \varepsilon_i .$$

We now calculate C_{2n}'. Let D_{2n}^+ be the set of ε in D_{2n} with $\Sigma\, \varepsilon_i \geq 0$, in other words at the most n terms equal to -1. The number of members of D_{2n}^+ is

$$\sum_{r=0}^{n} \binom{2n}{r} = 2^{2n-1} + \frac{1}{2}\binom{2n}{n} .$$

The total number of -1's occuring in these members is

$$\sum_{r=1}^{n} r\binom{2n}{r} = 2n \sum_{r=1}^{n} \binom{2n-1}{r-1} = 2n.2^{2n-2} = n2^{2n-1} .$$

By considering the difference between this and the number of +1's, we obtain

$$\frac{1}{2} C_{2n}' = \sum_{\varepsilon \in D_{2n}^+} \langle \varepsilon, e\rangle = 2n\left[2^{2n-1} + \frac{1}{2}\binom{2n}{n}\right] - n2^{2n}$$

$$= n\binom{2n}{n}.$$

The stated expressions for C_{2n-1}, C_{2n} and $\pi_1(\ell_1^n)$ follow. Also,

$C_{2n+2} = \left(1 + \frac{1}{2n}\right)C_{2n}$, so

$$\frac{K_{2n+2}}{K_{2n}} = \left(1 + \frac{1}{2n}\right)/\sqrt{1 + 1/n} > 1 ,$$

$$\frac{K_{2n+1}}{K_{2n-1}} = \frac{1}{2n} \sqrt{(2n+1)(2n-1)} < 1 .$$

The stated limit follows from the well-known "Wallis product".

We record the values for small n.

n	$\lambda(\ell_1^n)$	$\pi_1(\ell_1^n)$
2	1	2
3	3/2	2
4	3/2	8/3
5	15/8	8/3.

Before leaving this topic, let us observe that we have also evaluated $\pi_1^{(n)}(\ell_1)$:

<u>7.18.</u> $\pi_1^{(n)}(\ell_1) = \pi_1(\ell_1^n)$.

Proof. We show in fact that $\pi_1^{(n)}(\ell_1^p) = \pi_1^{(p)}(\ell_1^n)$. The statement then follows on letting $p \to \infty$. Given $a_1, \dots, a_n$ in ℓ_1^p, define $\tilde{a}_1, \dots, \tilde{a}_p$ in ℓ_1^n by : $\tilde{a}_j(i) = a_i(j)$. Then $\sum_i \|a_i\|_1 = \sum_j \|\tilde{a}_j\|_1$, and by 2.11,

$$\mu_1^{(1)}(\tilde{a}_1, \dots, \tilde{a}_p) = \mu_1^{(1)}(a_1, \dots, a_n) .$$

The assertion follows.

This enables us to strengthen 7.7 slightly:

<u>7.19 Corollary.</u> If E is any n-dimensional subspace of ℓ_1, then $d(E, \ell_\infty^n) \geq C_n$.

Proof. By 7.18, $\pi_1^{(n)}(E) \leq n/C_n$, while $\pi_1^{(n)}(\ell_\infty^n) = n$.

It was possible to perform this calculation using finite sums because of the existence of a finite norming set in the unit ball of $(\ell_1^n)^*$. This does not happen for complex ℓ_1^n , or for ℓ_2^n (real or complex). In such cases, one must expect to have to use integrals instead of finite sums. We consider this, and describe the evaluation of π_1 and λ for ℓ_2^n, in section 8.

König (1985) has given examples of n-dimensional spaces X_n with $\lambda(X_n)$ very close to $\sqrt{n}$. König and Lewis (not yet published) have also proved that no n-dimensional space has projection constant exactly equal to $\sqrt{n}$.

The dual of a C*-algebra

A C*-algebra is a sub-algebra of E of L(H) (for some Hilbert space H) such that T* is in E whenever T is in E. The next result, which is due to Tomczak-Jaegermann (1974), says that the dual of any C*-algebra (with identity) is of cotype 2, with $\kappa_2(E^*) \leqslant 2\sqrt{e}$. This can be regarded as a generalization of 7.6, since the diagonal operators on ℓ_2^n form a C*-algebra that is isometric to ℓ_∞^n ; its dual it therefore isometric to ℓ_1^n .

This result will not actually be deduced from Khinchin's inequality; rather, the proof is itself a very nice generalization of our second proof of the inequality. In contrast to our usual practice, we will give the proof for the complex case in the first instance.

We do not require much from the theory of operators on Hilbert spaces. We just need to recall that $\|T^*T\| = \|T\|^2$ for T in L(H). Consequently, if A,B are self-adjoint and AB = BA, then

$$\|A + iB\|^2 = \|(A - iB)(A + iB)\| = \|A^2 + B^2\| \qquad (1).$$

A functional ψ on E is said to be hermitian if $\psi(T^*) = \overline{\psi(T)}$ for all T (so that $\psi(A)$ is real for self-adjoint A).

7.20 Lemma. (i) If ψ is a hermitian functional on E, then

$$\|\psi\| = \sup\{\,|\psi(A)| : A \text{ self-adjoint and } \|A\| \leqslant 1\}.$$

(ii) Any functional ϕ on E is expressible as $\psi + i\chi$, where ψ,χ are hermitian and $\|\psi\|, \|\chi\| \leqslant \|\phi\|$.

Proof. (i) Take $\varepsilon > 0$. There exists T in E such that $\|T\| = 1$ and $\psi(T)$ is real and not less than $(1-\varepsilon)\|\psi\|$. Let $A = \frac{1}{2}(T+T^*)$. Then A is self-adjoint, $\|A\| \leqslant 1$ and $\psi(A) = \psi(T)$.

(ii) Define

$$\psi(T) = \tfrac{1}{2}\,\phi(T) + \tfrac{1}{2}\,\overline{\phi(T^*)}, \qquad i\chi(T) = \tfrac{1}{2}\,\phi(T) - \tfrac{1}{2}\,\overline{\phi(T^*)}\ .$$

7.21 Proposition. Let E be a complex C*-algebra containing the identity. Then $\kappa_2(E^*) \leqslant 2\sqrt{e}$.

Proof. We will show that for hermitian elements $\psi_1, \ldots ,\psi_n$ of E*,

$$(\sum_j \|\psi_j\|^2)^{\frac{1}{2}} \leqslant \sqrt{e}\ \rho_1(\psi_1, \dots ,\psi_n) .$$

Once this is done, the proof is completed as follows. Given arbitrary elements $\phi_1, \dots ,\phi_n$ of E^*, let $\phi_j = \psi_j + i\chi_j$ as in 7.20. Then $\|\phi_j\| \leqslant \|\psi_j\| + \|\chi_j\|$, so

$$(\sum_j \|\phi_j\|^2)^{\frac{1}{2}} \leqslant (\sum_j \|\psi_j\|^2)^{\frac{1}{2}} + (\sum_j \|\chi_j\|^2)^{\frac{1}{2}} .$$

Also, it is clear from the expression in 7.20 that $\| \sum \varepsilon_j\psi_j\| \leqslant \| \sum \varepsilon_j\phi_j\|$ for each ε in D_n, so $\rho_1(\psi_1, \dots ,\psi_n) \leqslant \rho_1(\phi_1, \dots ,\phi_n)$ (and similarly for χ_j).

Choose $\delta > 0$. For each j, there is a self-adjoint A_j in E such that $\|A_j\| = \|\psi_j\|$ and $\psi_j(A_j) \geqslant (1-\delta)\|\psi_j\|^2$. For ε in D_n, define $S_\varepsilon \in E$ by

$$S_\varepsilon = (I + i\varepsilon_1A_1) \dots (I + i\varepsilon_nA_n) .$$

By remark (1),

$$\|S_\varepsilon\|^2 \leqslant (1 + \|A_1\|^2) \dots (1 + \|A_n\|^2)$$
$$\leqslant \exp (\sum \|A_j\|^2).$$

Cancellation of the terms involving ε_j, $\varepsilon_j\varepsilon_k$, etc., shows that

$$\frac{1}{2^n} \sum_{\varepsilon \in D_n} \sum_j \varepsilon_j\psi_j(S_\varepsilon) = i \sum_j \psi_j(A_j) .$$

The absolute value of the left-hand side is not greater than

$$\frac{1}{2^n} \|S_\varepsilon\| \sum_j \|\varepsilon_j\psi_j\| \leqslant \exp \left(\tfrac{1}{2} \sum \|\psi_j\|^2\right) \rho_1(\psi_1, \dots ,\psi_n) .$$

Hence if $\sum \|\psi_j\|^2 = 1$, then

$$1-\delta \leqslant e^{\frac{1}{2}} \rho_1(\psi_1, \dots ,\psi_n).$$

This is true for all $\delta > 0$, so the statement follows.

This result does apply to the real case too, but only after some further discussion, as follows. Given a real Hilbert space H, one can define in an obvious way a complex Hilbert space H_C with elements $x + iy$ for $x,y \in H$, and $\|x + iy\|^2 = \|x\|^2 + \|y\|^2$. An operator T in L(H) extends automatically to give an operator in $L(H_C)$ with the same norm; we continue to denote the extension by T.

Let E be a C*-algebra in L(H), and let E_C be the set of operators in $L(H_C)$ of the form $A + iB$, where $A,B \in E$. Then E_C is a complex

C*-algebra. Clearly A + iB is self-adjoint if and only if $A^* = A$ and $B^* = -B$ (that is, B is "skew"). Any functional ϕ on E extends to a functional $\hat{\phi}$ on E_C in the obvious way : $\hat{\phi}(A + iB) = \phi(A) + i\phi(B)$.

A functional ψ on E is "symmetric" if $\psi(T^*) = \psi(T)$ for all T. Clearly, $\hat{\psi}$ is then hermitian. Also, both $\|\psi\|$ and $\|\hat{\psi}\|$ can be computed using self-adjoint operators as in 7.20, and $\psi(B) = 0$ for skew B. It follows that $\|\hat{\psi}\| = \|\psi\|$ (note: this does not apply to functionals on E in general!).

A functional χ on E is "anti-symmetric" if $\chi(T^*) = -\chi(T)$ for all T. It is easily verified that $i\hat{\chi}$ is then hermitian. Further, $\|\chi\| = \sup \{|\chi(B)| : B \text{ skew}, \|B\| \leqslant 1\}$. One deduces that $\|\hat{\chi}\| = \|\chi\|$.

Finally, any functional ϕ on E is expressible as $\psi + \chi$, where ψ is symmetric and χ anti-symmetric. The result now follows from the inequality for hermitian functionals proved in 7.21.

8. INTEGRAL METHODS ; GAUSSIAN AVERAGING

The basic results

As already mentioned, we can sometimes expect the exact evaluation of summing norms to require consideration of integrals rather than finite sums. In this section, we start by describing the (very simple) principles involved in doing this, and go on to apply them to the evaluation of $\pi_1(\ell_2^n)$, using ideas than can be thought of as a continuous version of the "averaging" of sections 6 and 7.

The reader who is unfamiliar with measure theory should not be put off by the first two results. Though they are stated in terms of measures, what we will actually use is a version requiring no more than ordinary Riemann integration on $\mathbb{R}^n$. For added clarity, we state the results for the case $p=1$ (which is how they will be applied). There should be no problem in formulation the corresponding statements for other p.

We have made repeated use of 3.2 : if we know of functionals f_j such that $\|Tx\| \leqslant \Sigma\, |f_j(x)|$ for all x, then $\pi_1(T) \leqslant \Sigma\, \|f_j\|$. The following is the natural generalization of this to a "continuous" instead of "discrete" collection of functionals. What is needed is a measure on the collection, allowing integration.

8.1. Let T be an operator on X. Suppose that there is a subset V of X^* and a positive measure μ on V such that

$$\|Tx\| \leqslant \int_V |f(x)|\, df \qquad \text{(with respect to } \mu)$$

for all $x \in X$. Then

$$\pi_1(T) \leqslant \int_V \|f\|\, df .$$

Proof. Let $\mu_1(x_1, \dots, x_k) = 1$, so that $\sum_i |f(x_i)| \leqslant \|f\|$ for each f. Then

$$\sum_i \|Tx_i\| \leqslant \sum_i \int_V |f(x_i)| \, df$$

$$\leqslant \int_V \|f\| \, df .$$

Typically, V will be either X^*, U_{X^*} or S_{X^*}.

Our second result, giving an inequality in the opposite sense, is a "continuous" analogue of the definition of π_1 itself. Given elements $x_1, \dots, x_k$ such that $\|f\| \geqslant \sum_i |f(x_i)|$ for all $f \in X^*$ (that is, $\mu_1(x_1, \dots, x_k) \leqslant 1$), the definition says that $\pi_1(T) \geqslant \sum_i \|Tx_i\|$. The analogous statement is :

<u>8.2.</u> Let T be an operator on a finite-dimensional space X. Suppose that there is a subset U of X and a positive measure μ on U such that

$$\|f\| \geqslant \int_U |f(x)| \, dx \qquad \text{(with respect to } \mu)$$

for all f in X^* (or a norming subset of U_{X^*}). Then

$$\pi_1(T) \geqslant \int_U \|Tx\| \, dx .$$

Proof. Take $\varepsilon > 0$. By 5.3, there exist elements f_j of X^* such that $\|Tx\| \leqslant \sum_j |f_j(x)|$ for all x, and $\sum_j \|f_j\| \leqslant (1 + \varepsilon)\pi_1(T)$. We have

$$\sum_j \|f_j\| \geqslant \sum_j \int_U |f_j(x)| \, dx$$

$$\geqslant \int_U \|Tx\| \, dx .$$

As remarked after 5.3, the functionals f_j can be chosen from a norming subset of U_{X^*} .

<u>Remarks.</u> (1) The promised version using only Riemann integration is as follows. Let X be $\mathbb{R}^n$ with some norm, and U a subset of X. Expressions like $\int_U h(x) \, dx$ are, of course, "multiple integrals" : x stands for $(x_1, \dots, x_n)$. The statement of 8.2 becomes : if there is a positive function

w on U such that $\|f\| \geqslant \int_U |f(x)|\, w(x)\, dx$ for all f, then $\pi_1(T) \geqslant \int_U \|Tx\|\, w(x)dx$. The statement of 8.1 translates similarly.

(2) The converse to 8.1 is essentially Pietsch's theorem, in combination with the Riesz representation theorem. The space X embeds into $C(U_{X^*})$, and there is a positive functional ϕ on $C(U_{X^*})$ such that $\|Tx\| \leqslant \phi(|x|)$ for all x and $\|\phi\| = \pi_1(T)$. There is a measure μ on U_{X^*} such that $\phi(|x|)$ equates to $\int |f(x)|\, df$. Using this in place of 5.3, one can show that 8.2 holds without the requirement that X should be finite dimensional.

(3) One can prove 8.2 directly, approximating to the integrals by finite sums. We sketch the argument for the case when U is bounded. Let U be the union of disjoint sets $U_1, \ldots ,U_k$ with diameters less than ε. Choose $u_i \in U_i$, and let $v_i = \mu(U_i)u_i$. Then $\Sigma\, |f(v_i)|$ approximates to $\int_U |f(x)|\, dx$ (hence $\mu_1(v_1, \ldots ,v_k) \leqslant 1+\varepsilon$) and $\Sigma\, \|Tv_i\|$ approximates to $\int_U \|Tx\|\, dx$.

Application to ℓ_2^2

Before attempting something more general, we illustrate the above ideas by applying them to (real) ℓ_2^2 . The elements with unit norm can of course be written as $y_\theta = (\cos\theta, \sin\theta)$ for $0 \leqslant \theta < 2\pi$. The norm is reproduced exactly by integration, as follows:

<u>8.3.</u> For x in ℓ_2^2 , $\|x\| = \frac{1}{4} \int_0^{2\pi} |\langle x,y_\theta\rangle|\, d\theta.$

Proof. Let $x = (r\cos\alpha, r\sin\alpha)$, so that $\|x\| = r$. Then $\langle x, y_\theta\rangle = r\cos(\theta - \alpha)$, which is positive for θ between $\alpha - \pi/2$ and $\alpha + \pi/2$. Now

$$\int_{\alpha-\pi/2}^{\alpha+\pi/2} \cos(\theta-\alpha)\, d\theta = 2.$$

It follows easily that

$$\int_0^{2\pi} |\langle x, y_\theta\rangle|\, d\theta = r \int_{\alpha-\pi}^{\alpha+\pi} |\cos(\theta-\alpha)|\, d\theta = 4r .$$

8.4. We have $\pi_1(\ell_2^2) = \pi/2$.

Proof. By 8.1,

$$\pi_1(\ell_2^2) \leq \frac{1}{4}\int_0^{2\pi} \|y_\theta\| \, d\theta = \pi/2 .$$

Since ℓ_2^2 coincides with its own dual, we can apply 8.2 with y_θ substituted for "x" to obtain the opposite inequality.

Exercise. Show by the steps indicated that if X is complex ℓ_1^2, then $\pi_1(X) = \pi/2$. Write $y_\theta = (1, e^{i\theta})$, and

$$g(x) = \int_{-\pi}^{\pi} |\langle x, y_\theta \rangle| \, d\theta .$$

Show that (i) $g(1,1) = 8$, (ii) $g(|x|) = g(x)$, (iii) $g(x_2, x_1) = g(x_1, x_2)$, (iv) $\|x\|_1 \leq \frac{1}{4} g(x)$. To apply 8.2, note that the functionals y_θ form a norming set and that $g(y_\theta) = 8$, by (i) and (ii).

Gaussian averaging

Recall that

$$\int e^{-\frac{1}{2}x^2} dx = \int x^2 e^{-\frac{1}{2}x^2} dx = \sqrt{2\pi} ,$$

both integrations being on the whole real line.

Write $G_1(x) = (1/\sqrt{2\pi}) e^{-\frac{1}{2}x^2}$ for $x \in \mathbb{R}$, and define the function G_n on $\mathbb{R}^n$ by:

$$G_n(x) = G_1(x_1) \dots G_1(x_n)$$

$$= \frac{1}{(2\pi)^{n/2}} e^{-\frac{1}{2}\|x\|^2} .$$

Here $x = (x_1, \dots, x_n)$ and $\| \ \|$ means $\| \ \|_2$. Clearly

$$\int_{\mathbb{R}^n} G_n(x) \, dx = \prod_{i=1}^{n} \int_{\mathbb{R}} G_1(x_i) \, dx_i = 1 .$$

In the following, we will be considering ordinary (Riemann) integrals on $\mathbb{R}^n$ with "weight" function G_n. Unless otherwise stated, the integration is over the whole of $\mathbb{R}^n$.

8.5. Let a,b be elements of $\mathbb{R}^n$. Then

$$\int \langle a,x\rangle \langle b,x\rangle G_n(x)\, dx = \langle a,b\rangle .$$

$$\int \langle a,x\rangle^2 G_n(x)\, dx = \|a\|_2^2 .$$

Proof. A bilinear form ϕ is defined on $\mathbb{R}^n$ by putting

$$\phi(a,b) = \int \langle a,x\rangle \langle b,x\rangle G_n(x)\, dx .$$

If $i \neq j$, then

$$\phi(e_i,e_j) = \int x_i x_j G_n(x)\, dx .$$

This equals 0, since $\int_{\mathbb{R}} x_i G_1(x_i)\, dx_i = 0$. Further,

$$\phi(e_i,e_i) = \int x_i^2 G_n(x)\, dx = 1 .$$

since $\int_{\mathbb{R}} x_i^2 G_1(x_i)\, dx_i = 1$. The first statement follows by bilinearity, and the second statement is obtained by taking $b = a$.

This is the continuous analogue of the finite averaging statements in 6.1. The 2^n elements of $\mathcal{E}$ of D_n have been replaced by the elements x of the whole of $\mathbb{R}^n$ (with the weighting factor $G_n(x)$).

Khinchin's inequality dealt with the average of $|\langle a,\varepsilon\rangle|$, showing that it lies between $(1/\sqrt{2})\|a\|_2$ and $\|a\|_2$. We now show that the "Gaussian" average of $|\langle a,x\rangle|$ does even better : it is a constant multiple of $\|a\|_2$ for all a. For the proof of this, we assume one major fact about integration on $\mathbb{R}^n$: it is "invariant under isometries." In other words, if f is integrable on $\mathbb{R}^n$ and T is an isometry of ℓ_2^n, then

$$\int f(Tx)\, dx = \int f(x)\, dx.$$

Since (clearly) $G_n(Tx) = G_n(x)$, it follows that

$$\int f(x)\, G_n(x)\, dx = \int f(Tx)\, G_n(x)\, dx.$$

8.6 Proposition. For any a in $\mathbb{R}^n$,

$$\int |\langle a,x\rangle|\, G_n(x)\, dx = \sqrt{2/\pi}\, \|a\|_2 .$$

Proof. We prove the statement first for the case $a = e_1$:

$$\int |\langle e_1,x\rangle| G_n(x)\, dx = \int_{\mathbb{R}} |x_1| G_1(x_1) dx_1 \int_{\mathbb{R}} G_2(x_2) dx_2 \dots \int_{\mathbb{R}} G_n(x_n) dx_n$$

$$= \int_{\mathbb{R}} |x_1|\, G_1(x_1)\, dx_1 \,.$$

$$= \frac{2}{\sqrt{2\pi}} \int_0^\infty x\, e^{-\frac{1}{2}x^2}\, dx$$

$$= \sqrt{2/\pi}\,.$$

Now let any a be given. There is an isometry T of ℓ_2^n such that $Ta = \|a\| e_1$. Then $\langle a,x\rangle = \langle Ta,Tx\rangle = \|a\|\langle e_1,Tx\rangle$, so

$$\int |\langle a,x\rangle|\, G_n(x)\, dx = \|a\| \int |\langle e_1,Tx\rangle|\, G_n(x)\, dx.$$

By the preceding remark, this equals $\|a\| \int |\langle e_1,x\rangle|\, G_n(x)\, dx$. We have just shown that the value of this is $\sqrt{2/\pi}$.

It now follows at once from 8.1 and 8.2 that (in the real case)

$$\pi_1(\ell_2^n) = \sqrt{\pi/2} \int \|x\|\, G_n(x)\, dx \,.$$

(For 8.1, we regard a as being in ℓ_2^n itself and the elements x as being in the dual, which of course also identifies with ℓ_2^n. For 8.2, we reverse this.) Before attempting an exact expression for this integral, we can show very easily that it already gives an improvement to the estimate in 7.11:

<u>8.7.</u> In the real case, $\pi_1(\ell_2^n) \leq \sqrt{n\pi/2}$ and $\lambda(\ell_2^n) \geq \sqrt{2n/\pi}$.

Proof. By 8.5, $\int x_1^2\, G_n(x)\, dx = 1$, and hence $\int \|x\|^2\, G_n(x)\, dx = n$. For any function f, we have $(\int fG_n)^2 \leq \int f^2 G_n$: this follows from $\int (f-c)^2 G_n \geq 0$, with c equal to $\int fG_n$. Hence $\int \|x\|\, G_n(x) dx \leq \sqrt{n}$, and the statments follow.

For an exact evaluation, we have to transform to n-dimensional polar co-ordinates $r,\theta_1, \dots ,\theta_{n-1}$. These are defined by:

$$x_1 = r \sin\theta_1 \sin\theta_1 \dots \sin\theta_{n-1},$$

$$x_k = r \cos\theta_{k-1} \sin\theta_k \dots \sin\theta_{n-1} \qquad (2 \leq k \leq n).$$

Here θ_1 ranges between 0 and 2π, and the other θ_j between 0 and π. The Jacobian is

$$r^{n-1} K(\theta_2, \dots, \theta_{n-1}) ,$$

where $K(\theta_2, \dots, \theta_{n-1}) = \sin\theta_2 \sin^2\theta_3 \dots \sin^{n-2}\theta_{n-1}$ (the exact expression for K does not matter for our purposes).

Write

$$I_n = \int_0^\infty x^n e^{-\frac{1}{2}x^2} dx .$$

Then $I_0 = \sqrt{\pi/2}$, $I_1 = 1$ and the value of I_n is deduced from the recurrence relation $I_n = (n-1) I_{n-2}$ for $n \geqslant 2$.

<u>8.8 Proposition.</u> (Gordon, 1969). Let $\pi_1(\ell_2^n) = B_n$ (real case). Then:

$$B_n = \sqrt{\pi/2}\,\frac{I_n}{I_{n-1}} = n \int_0^{\pi/2} \cos^n\theta \, d\theta ,$$

$$B_{2n} = \frac{1.3 \dots (2n-1)}{2.4 \dots (2n-2)}\,\frac{\pi}{2}, \quad B_{2n+1} = \frac{2.4 \dots (2n)}{1.3 \dots (2n-1)}.$$

$B_n/\sqrt{n}$ increases with n, and tends to $\sqrt{\pi/2}$ as $n \to \infty$.

Proof. Write $c_n = (2\pi)^{-n/2}$. We have

$$1 = \int G_n(x)dx = c_n \int_0^\infty r^{n-1} e^{-\frac{1}{2}r^2} dr \;(\textstyle\int K) ,$$

$$\int \|x\| \, G_n(x)dx = c_n \int_0^\infty r^n e^{-\frac{1}{2}r^2} dr \;(\textstyle\int K),$$

where $\int K$ means $\int K(\theta_2, \dots, \theta_{n-1}) \, d(\theta_2, \dots, \theta_{n-1})$ over the full range of these variables. Hence

$$\int \|x\| \, G_n(x) \, dx = I_n/I_{n-1} .$$

This gives the first expression for B_n. The other statements follow from the recurrence relation quoted and the well-known facts about $\int \cos^n\theta \, d\theta$ and the "Wallis" product.

<u>Notes.</u> (1) By 6.18, we have $\lambda(\ell_2^n) = n/B_n$.

(2) Comparison with 7.17 shows that $\pi_1(\ell_2^{2n+1}) = \pi_1(\ell_1^{2n+1})$ $(= B_{2n+1})$. The value of $\pi_1(\ell_1^{2n})$ is again B_{2n+1}, while $\pi_1(\ell_2^{2n})$ is smaller.

(3) Recall from 6.9 that $\pi_1^{(n)}(\ell_2^n) = \sqrt{n}$.

(4) The first integral in the proof shows that $\int K = (2\pi)^{n/2}/I_{n-1}$. Let f be a function defined on the sphere $S_n = \{x : \|x\|_2 = 1\}$. The "integral of f with respect to rotation-invariant measure" means $\int f(\underline{\theta})\, K(\underline{\theta})\, d\underline{\theta}$, where $\underline{\theta}$ stands for $(\theta_2, ..., \theta_{n-1})$. Usually such integrals are normalized by dividing by $\int K$ (so that the integral of 1 becomes 1). By transforming the integrals in 8.5 and 8.6 to polar form and considering the integration with respect to r (as we did in 8.8), one finds that these integrals translate to :

$$\int \langle a,x\rangle \langle b,x\rangle\, dx = \frac{1}{n} \langle a,b\rangle ,$$

$$\int |\langle a,x\rangle|\, dx = \frac{1}{B_n} \|a\|_2 ,$$

where these integrals are taken over S_n in the normalized form. (The reader may care to check the details).

We now show how to modify the above work for the complex case. Denote (temporarily) real and complex ℓ_2^n by $\ell_2^n(\mathbb{R})$, $\ell_2^n(\mathbb{C})$. Then $\ell_2^n(\mathbb{C})$ is isometric as a <u>real</u> space to $\ell_2^{2n}(\mathbb{R})$: the element $(x_1, x_2, ..., x_{2n})$ of $\ell_2^{2n}(\mathbb{R})$ corresponds to $(x_1+ix_2, ..., x_{2n-1}+ix_{2n})$. We denote both by x, but we must distinguish between the inner products in $\mathbb{C}$ and $\mathbb{R}^{2n}$: the notation $\langle\ \rangle$ will refer to the inner product of $\mathbb{C}^n$. Integrations will be over $\mathbb{R}^{2n}$.

The following is the complex equivalent of 8.6. Curiously, the constant is exactly the reciprocal of the one obtained in the real case.

<u>8.9</u> For a in $\mathbb{C}^n$,

$$\int |\langle a,x\rangle|\, G_{2n}(x)\, dx = \sqrt{\pi/2}\, \|a\|_2 .$$

Proof. The reduction to the case $a = e_1$ remains valid, but $|\langle e_1,x\rangle|$ now means $(x_1^2 + x_1^2)^{1/2}$. Hence

$$\int |\langle e_1,x\rangle|\, G_{2n}(x)\, dx = \int_{\mathbb{R}^2} (x_1^2 + x_2^2)^{1/2}\, G_1(x_1) G_2(x_2)\, d(x_1,x_2)$$

$$= \frac{1}{2\pi} \int_0^{2\pi} d\theta \int_0^{\infty} \rho^2 e^{-\frac{1}{2}\rho^2}\, d\rho$$

$$= \sqrt{\pi/2} .$$

8.10 Proposition. In the complex case,

$$\pi_1(\ell_2^n) = \frac{1.3 \dots (2n-1)}{2.4 \dots (2n-2)} = \frac{2}{\pi} B_{2n} .$$

$1/\sqrt{n}\ \pi_1(\ell_2^n)$ increases with n, and tends to $2/\sqrt{\pi}$ as $n \to \infty$.

Proof. By 8.1 and 8.2,

$$\pi_1(\ell_2^n) = \sqrt{\pi/2} \int \|x\|\ G_{2n}(x)\ dx = \frac{2}{\pi} B_{2n}.$$

(The inequality $\pi_1(\ell_2^n) \leqslant 2\sqrt{n/\pi}$ follows more easily, as in 8.7).

We finish this section with two further comments on Gaussian averaging. Firstly 8.6 (or 8.9) shows that ℓ_2^n embeds isometrically into the $L_1(\mu)$-space consisting of functions on $\mathbb{R}^n$ integrable in the sense that

$$\|f\|_1 = \int |f(x)|\ G_n(x)\ dx$$

is finite. Consequently, ℓ_2^n embeds "almost isometrically" into ℓ_1^N for suitable N (compare 7.9).

Secondly, the type 2 and cotype 2 constants have "Gaussian" variants (denoted by τ_2^G, κ_2^G) in which the discrete averages $\rho_2(a_1, \dots ,a_n)$ are replaced by

$$\gamma_2(a_1, \dots ,a_n) = \left[\int \|\sum_1 x_i a_i\|^2\ G_n(x)\ dx\right]^{1/2} .$$

The quantity γ_2 can be either greater or less than ρ_2, but it can be shown that

$$\sqrt{2/\pi}\ \tau_2(X) \leqslant \tau_2^G(X) \leqslant \tau_2(X),$$

$$\kappa_2^G(X) \leqslant \kappa_2(X) ,$$

(see [FDBS], lecture 19). Instead of relating the Gaussian constants to the ordinary ones, it is perhaps more profitable simply to regard them as giving an alternative parallel theory. For example, by applying 8.6 instead of Khinchin's inequality, we find (as in 7.6) that $\kappa_2^G(\ell_1) \leqslant \sqrt{\pi/2}$.

Exercise. For elements $a_1, \dots ,a_n$ of a Hilbert space, prove that $\gamma_2(a_1, \dots ,a_n)^2 = \Sigma\ \|a_i\|^2$.

9. 2-DOMINATED SPACES

Equivalent formulations of the property

It is a highly significant fact, first recognized by Grothendieck, that certain infinite-dimensional Banach spaces X have the property that all 2-summing operators on X are 1-summing, and that there is a constant K such that $\pi_1(S) \leq K\pi_2(S)$ for all such operators S. There is no generally agreed name for this property; we will adopt a term once used by Rosenthal and say that the space X is "2-dominated" if the above holds. We also write $\Delta_2(X)$ for the least constant K for which it holds (again, there is no standard notation; in Pietsch [OI], this quantity appears as $M_{2,1}(I_X)$).

Any finite-dimensional space is, of course, 2-dominated. For such spaces, the interest lies in determining the value of $\Delta_2(X)$. This can be regarded as another numerical parameter descriptive of the space (like $\lambda(X)$, $\kappa_2(X)$, etc.).

Trace duality provides an equivalent formulation of the property. Let α^*, β^* be dual norms to α, β under finite-dimensional trace duality. Restricting attention (for the moment) to finite-dimensional spaces, note that if we have $\beta(S) \leq K\alpha(S)$ for all S in L(X,Y), then $\alpha^*(T) \leq K\beta^*(T)$ for all T in L(Y,X). Since $\pi_2^* = \pi_2$ and $\nu_\infty^* = \pi_1$, we have that the following statements (for a particular finite-dimensional space X) are equivalent :

(i) for all finite-dimensional Y and all S in L(X,Y),

$$\pi_1(S) \leq K\pi_2(S) ;$$

(ii) for all finite-dimensional Y and all T in L(Y,X),

$$\pi_2(T) \leq K\nu_\infty(T) .$$

With a bit of care, we can show that there is no need for the restriction to finite-dimensional spaces here, and that weaker versions of both (i) and (ii) are also equivalent:

<u>9.1 Proposition.</u> Let X be any normed linear space. The

following statements are equivalent:

(i) for any normed linear space Y and any S in $P_2(X,Y)$, $\pi_1(S) \leqslant K\pi_2(S)$;

(i)' as (i), but with Y restricted to the spaces ℓ_2^k $(k \in \mathbb{N})$;

(ii) for any normed linear space Y and any T in FL(Y,X), $\pi_2(T) \leqslant K\nu_\infty(T)$;

(ii)' for any k and any T in $L(\ell_\infty^k, X)$, $\pi_2(T) \leqslant K\|T\|$.

Proof. It is enough to prove that (i)' implies (ii) and that (ii)' implies (i).

(i)' implies (ii). Take T as in (ii) and $\varepsilon > 0$. By 3.6, there exist k and A in $L(\ell_2^k,Y)$ such that $\|A\| = 1$ and $\pi_2(TA) \geqslant (1-\varepsilon)\pi_2(T)$. By 4.3, there exist S in $FL(X,\ell_2^k)$ with $\nu_2(S) \leqslant 1$ and trace $TAS \geqslant (1-\varepsilon)\pi_2(TA)$. By (i)', $\pi_1(S) \leqslant K\pi_2(S) \leqslant K$. By 4.2,

$$\begin{aligned}\text{trace } TAS = \text{trace } (AS)T &\leqslant \pi_1(AS)\, \nu_\infty(T)\\ &\leqslant \pi_1(S)\, \nu_\infty(T) \leqslant K\nu_\infty(T).\end{aligned}$$

So we have $(1-\varepsilon)^2\, \pi_2(T) \leqslant K\nu_\infty(T)$.

(ii)' implies (i). Take S as in (i) and $\varepsilon > 0$. There is a finite-dimensional subspace X_1 of X such that $\pi_1(S_1) \geqslant (1-\varepsilon)\,\pi_1(S)$, where $S_1 = S|_{X_1}$. We may assume that $S_1(X_1)$ is a subspace of some ℓ_∞^k. Trace duality applied to (ii)' (with X replaced by X_1) gives

$$\pi_1(S_1) \leqslant K\pi_2(S_1) \leqslant K\pi_2(S).$$

<u>Note.</u> An alternative proof that (ii)' implies (i) is as follows. Choose S and ε. By 3.6, there exist k and A in $L(\ell_\infty^k, X)$ such that $\|A\| = 1$ and $\pi_1(SA) \geqslant (1-\varepsilon)\,\pi_1(S)$. By (ii)', $\pi_2(A) \leqslant K$. By 5.12,

$$\pi_1(SA) \leqslant \pi_2(S)\, \pi_2(A) \leqslant K\pi_2(S).$$

It is easy to give direct proofs that (i)' implies (i) and (ii)' implies (ii) using the Pietsch factorization theorem and 4.11 respectively.

We now describe some elementary consequences of the definition.

<u>9.2.</u> If X_1 is a subspace of X, then $\Delta_2(X_1) \leqslant \Delta_2(X)$.

Proof. This is immediate from 9.1(ii)'. (It is also easy from formulation (i), using the extension theorem 5.9).

It is also clear that, for an infinite-dimensional space X, $\Delta_2(X)$ is the supremum of $\Delta_2(X_1)$ for finite-dimensional subspaces X_1 of X. Further, if Y is finitely represented in X (see section 0 for the meaning of this), then $\Delta_2(Y) \leq \Delta_2(X)$.

It is elementary that $\Delta_2(Y) \leq d(X,Y)\,\Delta_2(X)$. Also, if $d(Z, \ell_\infty^k) = C$ and T is in $L(X,Y)$, then $\pi_2(T) \leq C\Delta_2(X)\,\|T\|$. We now give a further equivalent form of (ii), in which ℓ_∞^k is replaced by any $\mathcal{L}_\infty$-space (in particular, $\ell_\infty(S)$ or $C(K)$).

<u>9.3.</u> Suppose that $\Delta_2(X) = K$. If Y is an $\mathcal{L}_\infty$-space and T is in $L(Y,X)$, then $\pi_2(T) \leq K\|T\|$.

Proof. Take $\varepsilon > 0$. There is a finite-dimensional subspace Y_1 of Y such that $\pi_2(T|Y_1) \geq (1-\varepsilon)\pi_2(T)$. There is a subspace Y_0, containing Y_1, such that $d(Y_0, \ell_\infty^k) \leq 1+\varepsilon$ for some k. Then $\pi_2(T_0) \leq (1+\varepsilon)K\,\|T_0\|$, where $T_0 = T|_{Y_0}$. The statement follows.

<u>9.4.</u> For any normed linear space X, we have $\pi_{2,1}(X) \leq \Delta_2(X)$. Hence a 2-dominated space has the Orlicz property.

Proof. Let $\mu_1(x_1, \ldots, x_k) = 1$. Then $\|T\| = 1$, where T is the operator in $L(\ell_\infty^n, X)$ such that $Te_i = x_i$ for each i. So $\pi_2(T) \leq \Delta_2(X)$. Clearly, $(\Sigma\,\|x_i\|^2)^{1/2} \leq \pi_2(T)$.

<u>9.5.</u> If dim X = n, then $\Delta_2(X) \leq \sqrt{n}$.

Proof. Let T be an operator from ℓ_∞^k to X. Then $T = I_X T$, so

$\pi_2(T) \leq \pi_2(I_X)\,\|T\| = \sqrt{n}\,\|T\|$.

<u>Exercise.</u> Prove 9.5 using formulation (i) of 9.1.

<u>9.6 Example.</u> $\Delta_2(\ell_\infty^n) = \sqrt{n}$, since for the identity in ℓ_∞^n we have $\|I\| = 1$, while $\pi_2(I) = \sqrt{n}$.

<u>9.7.</u> If dim X = n and $\Delta_2(X) = K$, then $\pi_1(X) \leq K\sqrt{n}$ and $\lambda(X) \geq \sqrt{n}/K$.

Proof. These inequalities follow at once from 9.1 (i),(ii) and the fact that $\pi_2(I_X) = \sqrt{n}$ (recall that $\lambda(X) = \nu_\infty(I_X)$).

This simple result betrays some of the strength of the 2-dominated property. Suppose that X is an infinite-dimensional space with the property. To within the constant multiple K ($= \Delta_2(X)$), 9.7 shows that every finite-dimensional subspace X_1 of X has $\pi_1(X_1)$ as small as possible and $\lambda(X_1)$ as large as possible.

Spaces with dual of type 2 ; Hilbert spaces

We now turn to the problem of establishing that certain spaces are 2-dominated. Khinchin's inequality, once again, is the key to our first result of this sort.

9.8 Proposition. Suppose that X^* is of type 2. Then X is 2-dominated, and $\Delta_2(X) \leqslant \sqrt{2}\, \tau_2(X^*)$.

Proof. For any subspace E of X, we have (by 6.12) $\tau_2(E^*) \leqslant \tau_2(X^*)$. Hence it is sufficient to prove the statement for finite-dimensional X. Take S in L(X,Y) and $\delta > 0$. By 5.3, there exist $f_1, \ldots, f_n$ in X^* such that

$$\|Sx\|^2 \leqslant \sum_i f_i(x)^2 .$$

for all x and

$$\Sigma \|f_i\|^2 \leqslant (1+\delta)^2\, \pi_2(S)^2 .$$

For $\varepsilon \in D_n$, write $g_\varepsilon = \sum_i \varepsilon_i f_i$. By Khinchin's inequality,

$$\|Sx\| \leqslant \frac{\sqrt{2}}{2^n} \sum_\varepsilon |g_\varepsilon(x)| \quad \text{for all x.}$$

Hence

$$\pi_1(S) \leqslant \frac{\sqrt{2}}{2^n} \sum_\varepsilon \|g_\varepsilon\| .$$

By the definition of τ_2, this is not greater than

$$\sqrt{2}\, \tau_2(X^*)\, (\sum_i \|f_i\|^2)^{\frac{1}{2}} ,$$

so we have

$$\pi_1(S) \leqslant \sqrt{2}\, \tau_2(X^*)(1+\delta)\, \pi_2(S).$$

Hence, in particular, $\Delta_2(H) \leqslant \sqrt{2}$ for any Hilbert space H. (Recall from 3.22 that for positive operators from ℓ_∞ to ℓ_2, it is elementary that $\pi_2(T) = \|T\|$). We return to a more precise evaluation of $\Delta_2(H)$ in a moment. Before doing this, we show that in this case, ℓ_∞ can be replaced by ℓ_1 in 9.1(ii)'.

9.9 Proposition. For any n,p and any S in $L(\ell_1^n, \ell_2^p)$, we have $\pi_2(S) \leqslant \Delta_2(H)\|S\|$. Similarly for S in $L(\ell_1, \ell_2)$.

Proof. Any S in $L(\ell_1^n, \ell_2^p)$ is of the form

$$Sx = \sum_{i=1}^{p} \langle a_i, x\rangle e_i ,$$

with the a_i in ℓ_∞^n. By 2.5, $\|S\| = \mu_2(a_1, \ldots, a_p)$. Let $\|S\| = 1$.
Any T in $L(\ell_\infty^n, \ell_2^m)$ is of the form

$$Ty = \sum_{j=1}^{m} \langle y, b_j\rangle e_j ,$$

with the b_j in ℓ_1^n, and $\|T\| = \mu_2(b_1, \ldots, b_m)$. Let $\|T\| = 1$. Then $\pi_2(T) \leqslant \Delta_2(H)$, and we have

$$\sum_j \|Sb_j\|^2 = \sum_i \sum_j \langle a_i, b_j\rangle^2 = \sum_i \|Ta_i\|^2 \leqslant \Delta_2(H)^2 .$$

The final statement follows in the usual way.

The reasoning in 9.9 is clearly reversible. We mention a nice application to infinite-dimensional theory:

9.10. Let X be a Banach space that contains an isomorphic copy of ℓ_1 (for example, ℓ_∞ or C(I)). Then there is a continuous (in fact, 2-summing) operator mapping X onto ℓ_2.

Proof. Let X_1 be the subspace isomorphic to ℓ_1. There is a continuous operator T_1 mapping X_1 onto ℓ_2. By 9.9, T_1 is 2-summing. By 5.9, it has a 2-summing extension defined on X.

Exercise. Let H be a Hilbert space, and let T be an operator in L(X,H). Use the Pietsch factorization theorem to show that $\pi_1(T) \leqslant \Delta_2(H)\, \pi_2(T^*)$.

We now return to the evaluation of $\Delta_2(H)$, using the Gaussian integrals introduced in section 8 (of course, the reader who is content with 9.8 can omit this). The method is basically the integral analogue of 9.8.

9.11 Proposition. $\Delta_2(\ell_2^n)$ tends to $\sqrt{\pi/2}$ (real case) or $2/\sqrt{\pi}$ (complex case) as $n \to \infty$.

Proof. Clearly $\Delta_2(\ell_2^n) \geq \pi_1(\ell_2^n)/\sqrt{n}$. By 8.8 and 8.10, this tends to the stated limits in the two cases.

To prove the reverse inequality, it is sufficient to prove that $\pi_1(S) \leq A\,\pi_2(S)$ for S in $L(\ell_2^n, \ell_2^m)$, where A is the appropriate constant. We give the details for the real case. Let $Sx = \sum_i \langle x,b_i\rangle e_i$. By 3.9, $\pi_2(S)^2 = \sum \|b_i\|^2$. By 8.6,

$$\|Sx\| = \sqrt{\pi/2} \int_{\mathbb{R}^m} | \sum_i y_i \langle x,b_i\rangle | \, G_m(y)\, dy .$$

By 8.1, it follows that

$$\pi_1(S) \leq \sqrt{\pi/2} \int \| \sum_i y_i b_i \| \, G_m(y)\, dy.$$

Recall from 8.5 that $\int y_i^2 \, G_m(y)\, dy = 1$ and $\int y_i y_j \, G_m(y)\, dy = 0$. Also, $(\int f G_m)^2 \leq \int f^2 G_m$ (see 8.7). Hence

$$\begin{aligned} \frac{2}{\pi}\, \pi_1(S)^2 &\leq \int \| \sum_i y_i b_i \|^2 \, G_m(y)\, dy \\ &= \int \sum_i y_i^2 \, \|b_i\|^2 G_m(y)\, dy \\ &= \sum_i \|b_i\|^2 \\ &= \pi_2(S)^2 . \end{aligned}$$

Exercise. By considering integrals on S_m instead of $\mathbb{R}^m$ (see note (4) following 8.8), show that in fact $\Delta_2(\ell_2^n) = \pi_1(\ell_2^n)/\sqrt{n}$.

Proposition 9.11, with the constants as stated, was obtained by Grothendieck (1956). The other results above, and the idea of recognizing the "2-dominated" property in its own right, can be largely traced to Dubinsky et al. (1972). In the next two sections, we will show that ℓ_1, and in fact all spaces of cotype 2, are 2-dominated. These are among the deepest results in this book.

Internal characterization of $\Delta_2(X)$

The following purely internal characterization is of interest, though it is not essential for our later results.

Given a finite sequence $(x_1, \dots, x_k)$ of elements of a normed linear space, define $\delta_2(x_1, \dots, x_k)$ to be the infimum of $(\Sigma\lambda_j^2)^{1/2}$ for (λ_j) such that $\|\Sigma t_j x_j\|^2 \leqslant \Sigma t_j^2 \lambda_j^2$ for all choices of (real) scalars t_j. (Again, the notation δ_2 is not standard; in [OI] the notation is $m_{(2,1)}$). Clearly:

(1) $\lambda_j \geqslant \|x_j\|$, so $\delta_2(x_1, \dots, x_k) \geqslant (\Sigma \|x_j\|^2)^{1/2}$.

(2) If $|t_j| \leqslant 1$ for all j, then $\|\Sigma t_j x_j\|^2 \leqslant \Sigma\lambda_j^2$, so

$$\delta_2(x_1, \dots, x_k) \geqslant \mu_1(x_1, \dots, x_k) .$$

9.12. Let T be an operator from ℓ_∞^k to X, and let $Te_j = x_j$. Then $\pi_2(T) = \delta_2(x_1, \dots, x_k)$.

Proof. (i) If (λ_j) satisfies (*), then for u in ℓ_∞^k,

$$\|Tu\|^2 = \|\sum_j u(j)x_j\|^2 \leqslant \sum_j \lambda_j^2 u(j)^2 .$$

Hence $\pi_2(T)^2 \leqslant \Sigma\, \lambda_j^2$. It follows that $\pi_2(T) \leqslant \delta_2(x_1, \dots, x_k)$.

(ii) By Pietsch's theorem, take λ_j such that $\Sigma\, \lambda_j^2 = \pi_2(T)^2$ and

$$\|Tu\|^2 = \|\sum_j u(j)x_j\|^2 \leqslant \sum_j \lambda_j^2 u(j)^2$$

for all u in ℓ_∞^k. Then (λ_j) satisfies (*), so $\delta_2(x_1, \dots, x_k) \leqslant \pi_2(T)$.

9.13 Corollary. The statement $\Delta_2(X) \leqslant K$ is equivalent to $\delta_2(x_1, \dots, x_k) \leqslant K\, \mu_1(x_1, \dots, x_k)$ for all finite sequences (x_j) in X.

Exercise. Prove the following statements :

(i) $\delta_2(x_1, \dots, x_k) = \inf\{(\Sigma\, \lambda_j^2)^{1/2} : x_j = \lambda_j y_j \text{ and } \mu_2(y_1, \dots, y_k) = 1\}$,

(ii) $\Sigma\, \|Sx_j\| \leqslant \pi_2(S)\, \delta_2(x_1, \dots, x_k)$.

Note that (ii) gives another alternative proof that (ii)' implies (i) in 9.1.

Exercise. Let $a_1, \dots, a_n$ be elements of a real Hilbert space with $\langle a_i, a_j\rangle \geqslant 0$ for all i,j. Write $\Sigma a_j = s$. Show that (*) is satisfied with $\lambda_j^2 = \langle a_j, s\rangle$, and hence that $\delta_2(a_1, \dots, a_n) = \|s\|$. Deduce that if $T : \ell_\infty^n \to H$ is such that $\langle Te_i, Te_j\rangle \geqslant 0$ for all i,j, then $\pi_2(T) = \|T\|$.

10. GROTHENDIECK'S INEQUALITY

Introduction

The result known as Grothendieck's inequality is coming to be recognized as one of the really major theorems of Banach space theory. It first appeared in Grothendieck (1956) under the title "the Fundamental theorem of the metric theory of tensor products" (as we have seen, a number of the other results considered in this book can be traced to the same memoir). In fact, the theorem admits a remarkable number of equivalent formulations, expressed variously in terms of summing norms, bilinear forms and tensor products. One version says that ℓ_1 has a property rather stronger than being 2-dominated. Some of these formulations were given by Grothendieck himself, others by later writers. A particularly elementary version was given by Lindenstrauss & Pelczynski (1968); this served to make the theorem much better known and understood by mathematicians generally.

The theorem has many applications, both within Banach space theory and in other areas, notably harmonic analysis (we cannot attempt to do justice to these in this book). Also, there is by now a repertoire of alternative proofs that must have few parallels in Mathematics. Despite this, the exact determination of the constant appearing in the inequality remains an unsolved problem. There are actually two separate problems, for the real and complex cases respectively.

In this section, we start with the Lindenstrauss-Pelczynski formulation, and give a version of the proof, due to Krivine (1979), that yields the best current estimate of the constant in the real case. We then derive several of the equivalent formulations. Finally, we describe a few of the more immediate applications. A second, completely different, proof is given in section 11, and some more applications follow in section 12.

The basic statement and Krivine's proof

We take as our "basic" version of Grothendieck's inequality the following statement, formulated by Lindenstrauss and Pelczynski (1968).

10.1 Theorem. There is a constant K_G (independent of m,n) such that the following holds. If $a_{i,j}$ $(1 \leqslant i \leqslant m,\ 1 \leqslant j \leqslant n)$ are real numbers such that $|\sum_i \sum_j a_{i,j} s_i t_j| \leqslant 1$ whenever $|s_i| \leqslant 1$, $|t_j| \leqslant 1$ for all i,j, then

$$|\sum_i \sum_j a_{i,j} \langle x_i, y_i \rangle| \leqslant K_G$$

whenever x_i, y_i are elements of the unit ball of a real Hilbert space. There is another constant $K_G^{\mathbb{C}}$ such that a similar statement holds for complex scalars.

The notation K_G, $K_G^{\mathbb{C}}$ is taken to mean the least possible constants in this statement. They are known as "Grothendieck's constants" for the real and complex case, and as already mentioned, their exact values have not been determined. If the Hilbert space is restricted to dimension n, the resulting constants are denoted by $K_G(n)$, $K_G^{\mathbb{C}}(n)$.

We will refer to the condition imposed on the $a_{i,j}$ in 10.1 as (LP).

Remark. It is equivalent, in the statement of 10.1, if we restrict to elements satisfying $\|x_i\| = \|y_i\| = 1$. For then if we are given elements with $\|x_i\| = \rho_i$, $\|y_i\| = \sigma_j$ (where $\rho_i, \sigma_j \leqslant 1$), it is clear that $b_{i,j}$ satisfies (LP), where $b_{i,j} = a_{i,j}\, \rho_i \sigma_j$, and we have

$$a_{i,j} \langle x_i, y_j \rangle = b_{i,j} \langle x_i', y_j' \rangle$$

where $\|x_i'\| = \|y_j'\| = 1$.

We now present, with some slight simplifications, the modification by Krivine (1979) of the proof of Lindenstrauss and Pelczynski for the real case. It yields the estimate $K_G \leqslant \pi/(2\log(1+\sqrt{2}))$ (= 1.782...). The starting point is the following geometrically motivated lemma on Gaussian integrals; the notation was introduced in section 8. By the "measure" of a set E in $\mathbb{R}^n$, we mean $\int_E G_n(x)\, dx$.

10.2 Lemma. Let u,v be elements of ℓ_2^n with $\|u\| = \|v\| = 1$. Then

$$\int_{\mathbb{R}^n} \operatorname{sign}\langle u,x\rangle \operatorname{sign}\langle v,x\rangle\, G_n(x)\, dx = \frac{2}{\pi} \sin^{-1}\langle u,v\rangle .$$

Proof. Let $\cos^{-1}\langle u,v\rangle = \theta$, so that $0 \leqslant \theta \leqslant \pi$ and $\sin^{-1}\langle u,v\rangle = \frac{\pi}{2} - \theta$.

Take α, β such that $0 < \beta - \alpha < \pi$, and let

$$E(\alpha,\beta) = \{(r\cos\theta,\ r\sin\theta,\ x_3, \dots ,x_n) : r \geqslant 0,\ \alpha \leqslant \theta \leqslant \beta\}.$$

The measure of $E(\alpha,\beta)$, as one would expect, is $(\beta-\alpha)/2\pi$, since after integrating with respect to $x_3, \dots ,x_n$, we are left with

$$\frac{1}{2\pi} \int_{E(\alpha,\beta)} e^{-\frac{1}{2}(x_1^2 + x_2)^2}\, dx_1\, dx_2 = \frac{1}{2\pi} \int_\alpha^\beta d\theta \int_0^\infty re^{-\frac{1}{2}r^2}\, dr.$$

Now consider the integral in question. There is an isometry T of ℓ_2^n such that $Tu = e_1$ and $Tv = (\cos\theta, \sin\theta, 0, \dots ,0)$.

The integral equals

$$\int_{\mathbb{R}^n} \operatorname{sign}\ \langle Tu,x\rangle \operatorname{sign}\ \langle Tv, x\rangle\, G_n(x)\, dx.$$

The set of x for which $\langle Tu,x\rangle > 0$ and $\langle Tv,x\rangle > 0$ is $E(\theta - \pi/2,\ \pi/2)$, which has measure $(\pi-\theta)/2\pi$ (a diagram helps !). The set of x for which $\langle Tu,x\rangle < 0$ and $\langle Tv,x\rangle > 0$ is $E(\pi/2,\ \theta + \pi/2)$, which has measure $\theta/2\pi$. Combining this with the negatives of these two sets, we see that the value of the integral is

$$2\ \frac{\pi - 2\theta}{2\pi} = 1 - \frac{2\theta}{\pi} = \frac{2}{\pi} \sin^{-1} \langle u,v\rangle .$$

10.3 Corollary. Suppose that $a_{i,j}$ satisfies (LP), and that u_i,v_j are elements of a Hilbert space with $\|u_i\| = \|v_j\| = 1$ for all i,j. Then

$$\Big| \sum_i \sum_j a_{i,j} \sin^{-1}\langle u_i v_j\rangle \Big| \leqslant \frac{\pi}{2} .$$

Proof. We may assume that the space is ℓ_2^n for some n. For each x in ℓ_2^n, condition (LP) gives

$$\Big| \sum_i \sum_j a_{i,j} \operatorname{sign}\ \langle u_i,x\rangle \operatorname{sign}\ \langle v_j,x\rangle \Big| \leqslant 1.$$

Taking the Gaussian integral over $\mathbb{R}^n$ and applying 10.2, we obtain the statement.

From this point, Lindenstrauss and Pelczynski finish the proof with a fairly short argument, obtaining Grothendieck's estimate $K_G \leq \sinh \pi/2$ (= 2.301..). We follow instead Krivine's method, which - at the cost of slightly more work - leads to the better estimate implied by $\sinh(\pi/2K_G) \geq 1$.

The point of the next lemma is to show that we can replace $\langle x,y \rangle$ by $\langle x,y \rangle^k$ in the basic statement. Recall from 6.1 that for x,y in $\mathbb{R}^n$,

$$\frac{1}{2^n} \sum_{\varepsilon \in D_n} \langle x,\varepsilon \rangle \langle \varepsilon,y \rangle = \langle x,y \rangle .$$

<u>10.4 Lemma.</u> For each positive integer k, there is a mapping $w_k : \ell_2^n \to \ell_2^N$ (where $N = 2^{nk}$) such that for all x,y,

$$\langle w_k(x), w_k(y) \rangle = \langle x,y \rangle^k .$$

Proof. Let H be the set of all real-valued mappings on the set $(D_n)^k$ (which has 2^{nk} elements), with inner product

$$\langle f,g \rangle_H = \frac{1}{2^{nk}} \sum \{f(e)g(e) : e \in (D_n)^k\} .$$

Given x in ℓ_2^n, let $w_{k,x}$ be the element of H given by

$$w_{k,x}(\varepsilon_1, \ldots ,\varepsilon_k) = \langle x,\varepsilon_1 \rangle \ldots \langle x,\varepsilon_k \rangle .$$

(here each ε_j is an element of D_n). It follows at once from 6.1 that

$$\langle w_{k,x} , w_{k,y} \rangle_H = \langle x,y \rangle^k .$$

<u>Notes.</u> (1) Putting y = x, we see that $\|w_k(x)\| = \|x\|^k$.

(2) One can use Gaussian integration (in particular, 8.5) instead of 6.1. The space ℓ_2^N is then replaced by the L_2 space corresponding to $\mathbb{R}^{kn}$ with Gaussian measure.

<u>10.5 Lemma.</u> Let c > 0. There are mappings u,v from ℓ_2^n to ℓ_2 such that for all x,y,

$$\langle u(x), v(y) \rangle = \sin c\langle x,y \rangle ,$$

$$\|u(x)\|^2 = \sinh (c\|x\|^2) , \quad \|v(y)\|^2 = \sinh (c\|y\|^2) .$$

Proof. Let $H_k = \ell_2^{N(k)}$, where $N(k) = 2^{nk}$, and let

$$H = (H_1 \times H_2 \times \ldots)_2 .$$

Of course, H is isometric to ℓ_2. Let I_k be the natural injection of H_k into H. With w_k as in 10.4, we have

$$\sin c\langle x,y\rangle = \sum_{k=1}^{\infty} (-1)^{k-1} c_k \langle w_{2k-1}(x), w_{2k-1}(y)\rangle$$

where $c_k = c^{2k-1}/(2k-1)!$. The required mappings are given by

$$u(x) = \sum_k \sqrt{c_k}\, I_k [w_{2k-1}(x)] ,$$

$$v(y) = \sum_k (-1)^{k-1} \sqrt{c_k}\, I_k [w_{2k-1}(y)] .$$

Proof of 10.1. As remarked after 10.1, we may assume that $\|x_i\| = \|y_j\| = 1$ for each i,j. Write $c = \sinh^{-1} 1 = \log(1 + \sqrt{2})$. With u,v as in 10.5, let $u_i = u(x_i)$, $v_j = v(y_j)$. Then $\|u_i\| = \|v_j\| = 1$, and

$$c\langle x_i,y_j\rangle = \sin^{-1}\langle u_i,v_j\rangle$$

(note that $|c\langle x_i,y_j\rangle| < 1$). So by 10.8,

$$\left| \sum_i \sum_j a_{i,j} \langle x_i,y_j\rangle \right| \leq \frac{\pi}{2c} .$$

Complex scalars

We now consider the relation between the real and complex cases. Of course we must distinguish between the real and complex version of condition (LP).

<u>10.6.</u> The real and complex versions of 10.1 imply each other, and $\frac{1}{2}K_G \leq K_G^{\mathbb{C}} \leq 2K_G$.

Proof. (i) Assume 10.1 for the real case, and suppose that the numbers $a_{j,k} = \lambda_{j,k} + i\mu_{j,k}$ satisfy complex-(LP). Then $\lambda_{j,k}$ and $\mu_{j,k}$ satisfy real-(LP). Let x_j,y_k be elements of the unit ball of a complex Hilbert space H, and let

$$S = \sum_j \sum_k a_{j,k} \langle x_j,y_k\rangle .$$

By multiplying by a suitable complex α, we may suppose that S is real. A real inner product is defined as H by : $\langle x,y\rangle_R = \mathrm{Re}\langle x,y\rangle$. Note that $\mathrm{Im}\langle x,y\rangle = \langle x,iy\rangle_R$. It follows that $|S| \leq 2\,K_G$.

(ii) Assume 10.1 for the complex case, and suppose that $a_{j,k}$ satisfy real-(LP). Let s_j, t_k be complex numbers with $|s_j|, |t_k| \leq 1$, and let

$$S' = \sum_j \sum_k a_{jk} s_j t_k.$$

Again, we may assume that S' is real, and it follows that $|S'| \leq 2$. Any real Hilbert space can be embedded in a complex one (in the way remarked after 7.21). Hence we obtain $K_G \leq 2 K_G^{\mathbb{C}}$.

Note. In (ii) we actually have $|S'| \leq K_G(2)$, since for complex s,t, one has $\mathrm{Re}\,(st) = \langle s, \bar{t} \rangle$, using the usual inner product on $\mathbb{R}^2$. Hence in fact $K_G \leq K_G(2) K_G^{\mathbb{C}}$. Krivine (1979) has shown that $K_G(2) = \sqrt{2}$ (though even this is not trivial).

One can prove directly that $K_G^{\mathbb{C}} \leq K_G$, but the best known estimates for $K_G^{\mathbb{C}}$ have been found by giving separate proofs specific to the complex case. Pisier (1978) showed that $K_G^{\mathbb{C}} \leq e^{1-\gamma}$ (= 1.5262..). Haagerup (1978) has improved the estimate to $K_G^{\mathbb{C}} \leq 1.4049...$. His method is roughly the complex analogue of Krivine's proof, but the technical details are much harder.

Equivalent formulations

We now describe some equivalent formulations of the statement. As we have already pointed out, there are a remarkable number of these, and each of them constitutes a significant theorem in its own right. The matrix $(a_{i,j})$ in 10.1 will be associated in turn with (i) a set of elements of $\mathbb{R}^n$, (ii) a bilinear form, (iii) an operator.

We start with a restatement in terms of summing norms. Naturally, this form is particularly relevant from the point of view of this book.

10.7 Theorem. Let H be a Hilbert space, and let T belong to $L(\ell_1^n, H)$ (for any n). Then $\pi_1(T) \leq K_G \|T\|$. Furthermore, this statement is equivalent to 10.1.

Proof. Let $a_1, \ldots, a_m$ be elements of ℓ_1^n. Then

$$\| \sum_i s_i a_i \| = \sum_j | \sum_i s_i a_i(j) | = \sum_j t_j \sum_i s_i a_i(j)$$

for suitable t_j with $|t_j| = 1$. Hence condition (LP) for the numbers $a_i(j)$ is equivalent to the statement $\mu_1(a_1, \ldots, a_m) \leq 1$.

Let T be in $L(\ell_1^n, H)$, and write $Te_j = y_j$. Then $\|T\| = \max \|y_j\|$. If $x_i \in H$ and $\|x_i\| \leq 1$ for each i, then

$$|\sum_i \langle x_i, Ta_i \rangle| \leq \sum_i \|Ta_i\| ,$$

and equality holds for suitable x_i. But $Ta_i = \sum_j a_i(j) y_j$, so

$$\sum_i \langle x_i, Ta_i \rangle = \sum_i \sum_j a_i(j) \langle x_i, y_j \rangle .$$

The equivalence of 10.1 and 10.7 (for both the real and complex case) is now clear.

Remarks. (1) The least constant in the statement of 10.7 is again K_G (or $K_G^{\mathbb{C}}$). Of course, $K_G(n)$ applies when dim H = n.

(2) In the usual way, ℓ_1^n can be replaced by ℓ_1 or any $\mathcal{L}_1$-space. In particular, all continuous operators from ℓ_1 to ℓ_2 are 1-summing, with $\pi_1(T) \leq K_G\|T\|$. (Of course, it follows that ℓ_1 is 2-dominated; we return to this point later).

(3) By trace duality, the following is also equivalent: for S in $L(\ell_2^m, \ell_1^n)$, $\nu_1(S) \leq K_G \nu_\infty(S)$.

10.8 Theorem. Given a bilinear form β on $\ell_\infty^m \times \ell_\infty^n$ and elements x_k of ℓ_∞^m, y_k of ℓ_∞^n $(1 \leq k \leq N)$, we have

$$|\sum_k \beta(x_k, y_k)| \leq K_G \|\beta\| \mu_2(x_1, \ldots, x_N) \mu_2(y_1, \ldots, y_N).$$

Furthermore, this statement is equivalent to 10.1.

Proof. As usual, we set out the proof for the real case; routine small modifications are required for the complex case.

A general bilinear form β is given by

$$\beta(x,y) = \sum_i \sum_j a_{i,j} x(i) y(j) ,$$

and condition (LP) is equivalent to $\|\beta\| \leq 1$. Given elements x_k of ℓ_∞^m and y_k of ℓ_∞^n, define elements $\tilde{x}_i$ $(1 \leq i \leq m)$ of ℓ_2^N by : $\tilde{x}_i(k) = x_k(i)$ (and $\tilde{y}_i$ similarly). By 2.6,

$$\mu_2(x_1, \ldots, x_N)^2 = \max_i \sum_k x_k(i)^2 = \max_i \|\tilde{x}_i\|^2,$$

so the condition $\mu_2(x_1, \ldots, x_N) \leq 1$ is equivalent to $\|\tilde{x}\| \leq 1$ for each i.

The equivalence with 10.1 is now clear from the equality

$$\sum_k \beta(x_k,y_k) = \sum_i \sum_j \sum_k a_{i,j}\, x_k(i)\, y_k(j)$$

$$= \sum_i \sum_j a_{i,j}\, \langle \tilde{x}_i, \tilde{y}_j \rangle .$$

Note that the Hilbert space has disappeared from this form of the statement. Clearly, $K_G(N)$ applies when the number of x_k, y_k is restricted to N.

Since only finite numbers of elements are involved, ℓ_∞^m and ℓ_∞^n can be replaced by any $\mathcal{L}_\infty$-spaces (in particular, C(S)).

We now prove a "Pietsch" type theorem for bilinear forms as in 10.8. This will provide us with a further equivalent version of Grothendieck's inequality in terms of bilinear forms. Again, we use notation appropriate to the real case.

10.9 Proposition. Let S,T be compact spaces, and let β be a bilinear form on $C(S) \times C(T)$. Then the following statements are equivalent:

(i) for all finite sequences (x_k, y_k) in $C(S) \times C(T)$,

$$| \sum_k \beta(x_k,y_k)| \leq \mu_2(x_1, \ldots ,x_N)\, \mu_2(y_1, \ldots ,y_N) ;$$

(ii) there exist positive functionals ϕ on C(S), ψ on C(T) such that $\|\phi\| = \|\psi\| = 1$ and $\beta(x,y)^2 \leq \phi(x^2)\psi(y^2)$ for all x,y.

Proof. First, assume (ii) and choose a sequence (x_k, y_k). Then

$$| \sum_k \beta\,(x_k,y_k) | \leq \sum_k \phi(x_k{}^2)^{1/2}\, \psi(y_k{}^2)^{1/2}$$

$$\leq (\Sigma\, \phi(x_k{}^2))^{1/2}\, (\Sigma\, \psi(y_k{}^2))^{1/2} \qquad \text{(by Schwarz's inequality)}$$

$$\leq \mu_2(x_1, \ldots ,x_N)\, \mu_2(y_1, \ldots ,y_N).$$

Now assume (i). It is enough to find ϕ,ψ such that

$$| \beta(x,y)| \leq \tfrac{1}{2}\, [\phi(x^2) + \psi(y^2)]$$

for all x,y. For then, if $\psi(y^2) > 0$, there exists λ such that $\lambda^2\psi(y^2) = \phi(x^2)$, and we have $|\beta(x,\lambda y| \leq \phi(x^2)$, hence $\beta(x,y)^2 \leq \phi(x^2)\, \psi(y^2)$.

Given x in C(S), let $\bar{x}$ be the function in C(S×T) defined by $\bar{x}(s,t) = x(s)$ (and define $\bar{y}$ similarly for y in C(T). Obviously, $\|\bar{x} + \bar{y}\| = \|x\| + \|y\|$ for positive x in C(S), y in C(T). Since $\mu_2(x_1, \ldots ,x_N)^2 = \| \Sigma x_i^2 \|$ and ab $\leq \frac{1}{2}(a^2+b^2)$ for positive a,b, the hypothesis implies that

$$| \Sigma \beta(x_k,y_k) | \leq \tfrac{1}{2} \| \Sigma x_k^2 \| + \tfrac{1}{2} \| \Sigma y_k^2 \| .$$

We will use Lemma 5.1. For a positive function h in C(S×T), let

$$q(h) = \sup\{ | \Sigma \beta(x_k,y_k) | : \tfrac{1}{2} \Sigma \bar{x}_k^2 + \tfrac{1}{2} \Sigma \bar{y}_k^2 \leq h\} .$$

Then q is clearly superlinear. It is clear from the above that $q(h) \leq \|h\|$. Hence there is a positive functional F on C(S×T) such that $\|F\| \leq 1$ and $F(h) \geq q(h)$ for all $h \geq 0$. For x in C(S), y in C(T), we have

$$|\beta(x,y)| \leq \tfrac{1}{2} F(\bar{x}^2 + \bar{y}^2) = \tfrac{1}{2} \phi(x^2) + \tfrac{1}{2} \psi(y^2),$$

as required.

So we can now state the following, which is one of the most frequently quoted versions of Grothendieck's inequality.

<u>10.10 Theorem.</u> Let S,T be compact spaces, and let β be a bilinear form on C(S) × C(T) (real or complex). Then there exist positive functionals ϕ on C(S), ψ on C(T) such that $\|\phi\| = \|\psi\| = 1$ and

$$|\beta(x,y)| \leq K_G \|\beta\| \phi(|x|^2)^{\frac{1}{2}} \psi(|y|^2)^{\frac{1}{2}}$$

for all x,y. Furthermore, this statement is equivalent to 10.1.

Our next reformulation is an easy step from 10.8. It uses the following variant of 2-summing norm. Let Y be $\mathbb{R}^n$ with any lattice norm. As in section 7, we consider elements of the form $(\Sigma y_k^2)^{\frac{1}{2}}$. For T in L(X,Y), we define

$$\tilde{\pi}_2(T) = \sup \{ \| [\Sigma(Tx_k)^2]^{\frac{1}{2}} \| : \mu_2(x_1, \ldots ,x_N) \leq 1\}.$$

One can verify that $\tilde{\pi}_2$ is a norm (though this is not important for our purposes). Our interest is in the case when X is ℓ_∞^m (or ℓ_∞) and Y is ℓ_1^n (or ℓ_1). Note that for this case it follows from 7.5 that $\tilde{\pi}_2(T) \geq \pi_2(T)$.

<u>10.11 Lemma.</u> Let $y_1, \dots, y_N \in \ell_1^n$, and let $y = (\Sigma\, y_k^2)^{1/2}$. Then

$$\|y\| = \sup \{ \sum_k \langle y_k, v_k \rangle : v_k \in \ell_\infty^N \text{ and } \mu_2(v_1, \dots, v_N) \leq 1 \}.$$

Proof. Let $\mu_2(v_1, \dots, v_N) \leq 1$, so $\sum_k v_k(j)^2 \leq 1$ for each j. Then $\sum_k y_k(j)\, v_k(j) \leq y(j)$, by Schwarz's inequality, so

$$\sum_k \langle y_k, v_k \rangle \leq \sum_j y(j) = \|y\| .$$

Now define $v_k(j)$ to be $y_k(j)/y(j)$ if $y(j) \neq 0$, and 0 otherwise. Then $\sum_k v_k(j)^2 \leq 1$ and $\sum_k y_k(j) v_k(j) = y(j)$ for each j. Hence $\sum_k \langle y_k, v_k \rangle = \|y\|$.

<u>10.12 Theorem.</u> If T is in $L(\ell_\infty^m, \ell_1^n)$ (for any m,n), then $\pi_2(T) \leq K_G \|T\|$. Furthermore, this statement is equivalent to 10.1.

Proof. There is an isometry between $B(\ell_\infty^m, \ell_\infty^n)$ (the space of bilinear forms) and $L(\ell_\infty^m, \ell_1^n)$, defined by : $\beta(u,v) = \langle Tu, v \rangle$. By the lemma, it is now clear that the statements in 10.8 and 10.12 are equivalent.

It is easily checked that this applies in the complex case too.

Again, ℓ_∞^m can be replaced by any $\mathcal{L}_\infty$-space, and ℓ_1^n by any $\mathcal{L}_1$-space. The constant is $K_G(N)$ when the number of elements x_i is restricted to N.

This version (10.12) is the form used by our second proof of Grothendieck's inequality (see section 11). It can be generalized to operators between any pair of Banach lattices (see [CBS II], 1f).

<u>Exercise.</u> Give a direct proof of the equivalence of 10.12 and 10.7 (write the operators concerned as $\Sigma \langle a_i, x \rangle e_i$ and $\Sigma \langle b_j, y \rangle e_j$).

Our final reformulation, which is one of Grothendieck's original ones, is seemingly quite different. It dispenses with the $a_{i,j}$ entirely, and takes the form of an assertion about tensor expressions for inner products.

Let S,T be compact spaces. The algebraic tensor product $C(S) \otimes C(T)$ can be identified with a subspace of $C(S \times T)$ by taking $f \otimes g$ to be the function $h(s,t) = f(s)g(t)$. For present purposes, this may be regarded as the definition of $C(S) \otimes C(T)$. The "projective" norm γ is defined by :

$$\gamma(u) = \inf \{ \Sigma \|f_r\| . \|g_r\| : u = \Sigma f_r \otimes g_r \} .$$

Recall from 1.19 that the dual of $C(S) \otimes_\gamma C(T)$ identifies with the space of bilinear forms on $C(S) \times C(T)$ (our new interpretation of $C(S) \otimes C(T)$ makes no difference).

10.13 Theorem. Let $S_n = \{x \in \ell_2^n : \|x\| = 1\}$, and let P_n be the inner-product function on $S_n \times S_n$, that is : $P_n(x,y) = \langle x,y \rangle$. Then $\gamma(P_n) = K_G(n)$ (so tends to K_G as $n \to \infty$).

Proof. Suppose that $\langle x,y \rangle = \sum_r f_r(x) g_r(y)$. Let $(a_{i,j})$ satisfy (LP) and let x_i, y_j be elements of S_n. For each r,

$$\left| \sum_i \sum_j a_{i,j} \, f_r(x_i) \, g_r(y_j) \right| \leq \|f_r\| . \|g_r\|.$$

Summation over r gives

$$\left| \sum_i \sum_j a_{i,j} \langle x_i, y_j \rangle \right| \leq \sum_r \|f_r\| . \|g_r\| ,$$

and hence $K_G(n) \leq \gamma(P_n)$.

To prove the converse, define δ_i in $C(S_n)$ by : $\delta_i(x) = x(0)$. Then $P_n = \sum_i \delta_i \otimes \delta_i$, and $\sum_i \delta_i(x)^2 = 1$ for all x in S_n, so $\mu_2(\delta_1, \ldots, \delta_n) = 1$. By 1.19, there is a bilinear form β on $C(S_n) \times C(S_n)$ such that $\|\beta\| = 1$ and $\gamma(P_n) = \sum_i \beta(\delta_i, \delta_i)$. By 10.8, the value of this is not greater than $K_G(n)$.

Lower bounds for K_G

10.14. $K_G \geq \pi/2$, while $K_G^C \geq 4/\pi$.

Proof. Recall from 9.11 that $\Delta_2(\ell_2^n) \to A$ as $n \to \infty$, where A is $\sqrt{\pi/2}$ in the real case, $2/\sqrt{\pi}$ in the complex case. This means that for any $\varepsilon > 0$, there exist m,n and an operator T from ℓ_∞^m to ℓ_2^n such that $\|T\| = 1$ and $\pi_2(T) \geq (1-\varepsilon)A$. So there are elements x_k with $\mu_2(x_1, \ldots, x_N) = 1$ and $\sum \|Tx_k\|^2 \geq (1-\varepsilon)^2 A^2$. Define β on $\ell_\infty^m \times \ell_\infty^m$ by : $\beta(x,y) = \langle Tx, Ty \rangle$. Then $\|\beta\| = 1$ and $\sum \beta(x_k, x_k) \geq (1-\varepsilon)^2 A^2$. The statements follow, by 10.8. (In the complex case, β is actually "sesquilinear", not bilinear, but it is easily checked that 10.8 still applies).

These lower bounds were stated by Grothendieck. Krivine and A.M. Davie (unpublished) have improved them to : $K_G \geq 1.676\ldots$, $K_G^C \geq 1.338\ldots$.

ℓ_1 is 2-dominated

Both 10.7 and 10.12 imply that ℓ_1 is 2-dominated, using the appropriate alternative formulations of this property from 9.1. Clearly, $\Delta_2(\ell_1) \leqslant K_G$. Like K_G itself, the exact value of $\Delta_2(\ell_1)$ is not known, and one can expect it to be different in the real and complex cases. Denote it temporarily by K_1. In the real case, we have $K_1 \geqslant \sqrt{2}$, since $\pi_2(I_{1,2}) = 1$, while $\pi_1(I_{1,2}) = \sqrt{2}$ (6.3, 7.10); in fact, this applies to ℓ_1^2 as well as ℓ_1.

Of course, if X is any space that is finitely represented in ℓ_1 (e.g. $C(S)^*$; see 0.24), then $\Delta_2(X) \leqslant K_1$.

Recall that for T in $L(\ell_1,\ell_2)$, we have $\pi_2(T) \leqslant A\|T\|$, where A is $\sqrt{\pi/2}$ in the real case, $2/\sqrt{\pi}$ in the complex case (9.9, 9.11). Hence $K_G \leqslant AK_1$, and any direct proof that ℓ_1 is 2-dominated affords an alternative proof of Grothendieck's inequality (in the form 10.7). It is instructive to see how one can also obtain the form 10.10, again with AK_1 in place of K_G.

Proof of a version of 10.10. Assume $\|\beta\| = 1$. Define $B : C(S) \to C(T)^*$ by $(Bx)(y) = \beta(x,y)$. Then $\|B\| = 1$, so $\pi_2(B) \leqslant K_1$. By 5.8, there exist a Hilbert space H and $B_1 : C(S) \to H$, $B_2 : H \to C(T)^*$ such that $B = B_2B_1$ and $\pi_2(B_1) \leqslant K_1$, $\|B_2\| = 1$. Let $\hat{B}_2$ be the restriction to C(T) of $B_2^* : C(T)^{**} \to H$. Since $\Delta_2(H) = A$, we have $\pi_2(\hat{B}_2) \leqslant A$. Now

$$\beta(x,y) = (B_2B_1x)(y) = (\hat{B}_2y)(B_1x) .$$

By Pietsch's theorem, there are positive functionals ϕ on C(S), ψ on C(T) such that $\|\phi\| = \|\psi\| = 1$ and $\|B_1x\|^2 \leqslant K_1^2\phi(x^2)$, $\|\hat{B}_2y\|^2 \leqslant A^2\psi(y^2)$. Hence $|\beta(x,y)| \leqslant AK_1\ \phi(x^2)^{1/2}\ \psi(y^2)^{1/2}$.

Two further applications

To exemplify what can be done with Grothendieck's inequality, we mention two simple applications to Banach-Mazur distances and projection constants. We continue to distinguish K_G and K_1.

<u>10.15</u> (Gordon, 1968). For any n-dimensional space X,

$$\pi_1(X) \leqslant K_G\ d(X,\ell_1^n)\, d(X,\ell_2^n) .$$

Proof. There are operators $S : \ell_1^n \to X$ and $T : X \to \ell_2^n$ such that $\|S\| = \|T\| = 1$ and $\|S^{-1}\| = d(X,\ell_1^n)$, $\|T^{-1}\| = d(X,\ell_2^n)$. By 10.7, $\pi_1(TS) \leqslant K_G$. Now $I_X = T^{-1}(TS)S^{-1}$, so $\pi_1(I_X) \leqslant K_G\|T^{-1}\|.\|S^{-1}\|$.

Of course, we knew already that $d(X,\ell_1^n)\,d(X,\ell_2^n) \geq d(\ell_1^n,\ell_2^n) = \sqrt{n}$.

<u>10.16</u> (Garling & Gordon, 1971). For any n-dimensional space X, we have $\lambda(X)\lambda(X^*) \geq \sqrt{n}/K_1$.

Proof. We may assume that X is a subspace of some ℓ_∞^k. Let P be a projection of ℓ_∞^k onto X with $\|P\| = \lambda(X)$. Then $P^*(X^*)$ is a subspace of ℓ_1^k, so by 9.7, $\lambda[P^*(X^*)] \geq \sqrt{n}/K_1$. For f in X^*, it is clear that P^*f is an extension of f, so $\|P^*f\| \geq \|f\|$. Hence

$$d[X^*,\ P^*(X^*)] \leq \|P^*\| = \lambda(X)\ ,$$

so that

$$\frac{\sqrt{n}}{K_1} \leq \lambda[P^*(X^*)] \leq \lambda(X^*)\ d[X^*,P^*(X^*)] \leq \lambda(X)\lambda(X^*)\ .$$

Note that ℓ_1^n is itself an example of a space X for which $\lambda(X)\lambda(X^*) < \sqrt{n}$.

The Grothendieck property

It is natural to ask whether other spaces could replace ℓ_1 in Grothendieck's inequality as formulated in 10.7. A space X is said to have the "Grothendieck property" with constant K if for all operators T from X to a Hilbert space, we have $\pi_1(T) \leq K\|T\|$. Note that this implies that X is 2-dominated, with $\Delta_2(X) \leq K$ (see 9.1). However, ℓ_2 itself does not have the Grothendieck property (consider the identity !), although it is 2-dominated. Further, while the 2-dominated property is inherited by subspaces, this is not the case for the Grothendieck property (recall that ℓ_2^n embeds into ℓ_1). What can be said in this direction is the following:

<u>10.17.</u> If X has the Grothendieck property with constant K, and X_1 is a complemented subspace of X (with projection P), then X_1 has the Grothendieck property with constant $K\|P\|$.

Proof. Let T be an operator from X_1 to a Hilbert space H. Then TP is in L(X,H), so $\pi_1(TP) \leq K\|TP\| \leq K\|P\|.\|T\|$. Since TP is an extension of T, we have $\pi_1(T) \leq \pi_1(TP)$.

We will see in Section 12 that a space with a "good" basis and the

Grothendieck property is isomorphic to ℓ_1^n (or ℓ_1). However, there are spaces with the property that are not $\mathfrak{L}_1$-spaces. For a good survey of results related to this, see [FLO], Chapter 6.

Exercise. Prove that X has the Grothendieck property with constant K if and only if for all S in $L(X^*, \ell_1^n)$, $\pi_2(S) \leqslant K'\|S\|$ (where $K' \leqslant K \leqslant K'\Delta_2(H)$).

(Necessity can be proved directly. For sufficiency, restate the Grothendieck property using trace duality, and apply 4.11.)

Concluding remarks

As observed earlier, there are many further applications of Grothendieck's inequality. We will outline a few of them in Section 12, in conjunction with basis constants.

A second proof of the inequality (giving version 10.12) is described in Section 11. This has the attraction of being free of integration, and it adapts to show that all cotype-2 spaces are 2-dominated. However, it gives a less accurate estimate of the constant.

Among the many other methods of proof that have been devised, we mention one that is at first sight very appealing. Suppose that we can find a Banach space X and a 1-summing operator Q of X onto ℓ_2 (so that Q is M-open for some M). Let T be in $L(\ell_1, \ell_2)$. Then there is an operator T : $\ell_1 \to X$ such that $T = QT_1$ and $\|T_1\| \leqslant M\|T\|$ (this is elementary). Hence

$$\pi_1(T) \leqslant \pi_1(Q)\|T_1\| \leqslant M\pi_1(Q)\|T\|,$$

so version 10.7 of the statement follows. However, no entirely easy way is known of exhibiting such X and Q, or finite-dimensional equivalents (see [FLO], Chapter 5).

11. THE INTERPOLATION METHOD FOR GROTHENDIECK-TYPE THEOREMS

An interpolation theorem

We now prove a theorem that, for operators on ℓ_∞^n, establishes a relationship between $\pi_p(T)$ for different p. For this purpose, we relax our general principle of concentrating on the cases p = 1,2. In fact, our applications of the theorem (including the promised alternative proof of Grothendieck's inequality) will make use of the statement with "r" equal to 4. The proof is a direct application of Pietsch's theorem, together with Hölder's inequality in the form : $\Sigma\, \alpha_j^{\theta}\beta_j^{1-\theta} \leqslant (\Sigma\, \alpha_j)^{\theta}(\Sigma\, \beta_j)^{1-\theta}$.

11.1 Theorem. Let T be any operator from ℓ_∞^n to a normed linear space Y. Let $1 \leqslant p < r$, and write $\frac{p}{r} = \theta$. Then

$$\pi_r(T) \leqslant \pi_p(T)^{\theta}\, \|T\|^{1-\theta}.$$

In particular, $\pi_{2p}(T)^2 \leqslant \pi_p(T)\, \|T\|$.

Proof. Write $Te_j = a_j$. By Pietsch's theorem (5.2), there is a positive linear functional ϕ on ℓ_∞^n such that $\|\phi\| = \pi_p(T)^p$ and $\|Tx\|^p \leqslant \phi(|x|^p)$ for all x. For certain $\alpha_j \geqslant 0$, we have $\phi(x) = \Sigma\, \alpha_j\lambda_j$, where $x = (\lambda_1, \dots, \lambda_n)$. Then $Tx = \Sigma\, \lambda_j a_j$, and the above conditions become

$$\Sigma\, \alpha_j = \pi_p(T)^p\,, \tag{1}$$

$$\|\Sigma\, \lambda_j a_j\|^p \leqslant \Sigma\, \alpha_j|\lambda_j|^p. \tag{2}$$

Choose $x = (\lambda_1, \dots, \lambda_n)$. There is an element f of U_{Y^*} such that $f(\Sigma\, \lambda_j a_j) = \|\Sigma\, \lambda_j a_j\|$. Write $|f(a_j)| = \beta_j$. Then $\beta_j = \varepsilon_j f(a_j)$, where $|\varepsilon_j| = 1$, so

$$\Sigma\, \beta_j = f(\Sigma\, \varepsilon_j a_j) = f\,[T(\varepsilon_1,\dots,\varepsilon_n)] \leqslant \|T\|\,. \tag{3}$$

Also, for $\mu_j \geqslant 0$,

$$\Sigma\, \mu_j \beta_j = f(\Sigma\, \varepsilon_j \mu_j a_j)\ ,$$

so

$$(\Sigma\, \mu_j \beta_j)^p \leqslant \|\Sigma\, \varepsilon_j \mu_j a_j\|^p \leqslant \Sigma\, \alpha_j \mu_j^p\ , \tag{4}$$

by (2). By Hölder's inequality,

$$\begin{aligned}\|\Sigma\, \lambda_j a_j\| &= f(\Sigma\, \lambda_j a_j) \leqslant \Sigma\, |\lambda_j| \beta_j \\ &= \Sigma(|\lambda_j| \beta_j^{\theta})\, \beta_j^{1-\theta} \\ &\leqslant (\Sigma\, \mu_j \beta_j)^{\theta}\, (\Sigma\, \beta_j)^{1-\theta}\end{aligned}$$

where $\mu_j = |\lambda_j|^{1/\theta}$. Remembering that $r\theta = p$, we now have from (3) and (4):

$$\|\Sigma\, \lambda_j a_j\|^r \leqslant (\Sigma\, \alpha_j \mu_j^p)\, \|T\|^{r-p} = \|T\|^{r-p}\, \Sigma\, \alpha_j |\lambda_j|^r\ .$$

By 3.17 (the easy converse of Pietsch's theorem), this implies that

$$\begin{aligned}\pi_r(T)^r &\leqslant \|T\|^{r-p}(\Sigma\, \alpha_j) \\ &= \|T\|^{r-p}\, \pi_p(T)^p\ .\end{aligned}$$

This is equivalent to the statement.

In particular, $\pi_2(T)^2 \leqslant \pi_1(T)\, \|T\|$ (the above proof becomes rather simpler for this case). Easy examples show that this does not hold for operators defined on spaces other than ℓ_∞^n : in fact, when applied to I_X (where $\dim X = n$), it gives $\pi_1(X) \geqslant n$. As we know, this fails for ℓ_1^n, ℓ_2^n. An immediate consequence is:

<u>11.2.</u> For any finite-dimensional X, we have $\Delta_2(X)^2 \leqslant \pi_1(X)$.

Proof. For any operator from ℓ_∞^n to X, we have $\pi_1(T) \leqslant \|T\|\pi_1(X)$, and hence $\pi_2(T)^2 \leqslant \pi_1(X)\, \|T\|^2$.

For T as in 11.1, it can also be shown that

$$\pi_r(T) \leqslant \pi_p(T)^{\theta} \pi_q(T)^{1-\theta}\ ,$$

where $p < r < q$ and $\frac{1}{r} = \frac{\theta}{p} + \frac{1-\theta}{q}$.

Second proof of Grothendieck's inequality

We prove Grothendieck's inequality in the form 10.12, using 11.1 and Khinchin's inequality. The proof adapts easily for complex scalars. Recall from 10.12 the meaning of $\tilde{\pi}_2(T)$.

11.3 Lemma. Let Y be $\mathbb{R}^n$ with any lattice norm, and let T be an operator from ℓ_∞^n to Y. Then

$$\tilde{\pi}_2(T) \leq 3^{1/4}\, 2^{1/2}\, \pi_4(T).$$

Proof. Take elements $x_1, \dots, x_k$ of ℓ_∞^n. For ε in D_k, write $y_\varepsilon = \Sigma\, \varepsilon_i x_i$. By Khinchin's inequality, applied pointwise,

$$[\Sigma\, (Tx_i)^2]^{1/2} \leq \frac{\sqrt{2}}{2^k} \sum_\varepsilon |Ty_\varepsilon| . \tag{1}$$

By 7.2, also applied pointwise,

$$\frac{1}{2^k} \sum_\varepsilon y_\varepsilon^4 \leq 3\, (\Sigma\, x_i^2)^2 . \tag{2}$$

Recall that $\frac{1}{N}(c_1+\dots+c_N) \leq [\frac{1}{N}(c_1^4+\dots+c_N^4)]^{1/4}$ for $c_i \geq 0$. Using this and the definition of π_4 for operators on ℓ_∞^m, we have :

$$\begin{aligned} \|\,[\Sigma\,(Tx_i)^2]^{1/2}\| &\leq \frac{\sqrt{2}}{2^k} \sum_\varepsilon \|Ty_\varepsilon\| \qquad \text{(by (1))} \\ &\leq \sqrt{2}\, [\frac{1}{2^k}\, \Sigma\, \|Ty_\varepsilon\|^4]^{1/4} \\ &\leq \sqrt{2}\, \pi_4(T)\, \|\frac{1}{2^k}\, \Sigma\, y_\varepsilon^4\|^{1/4} . \\ &\leq \sqrt{2}\, \pi_4(T)\, 3^{1/4}\, \|\Sigma\, x_i^2\|^{1/2} \qquad \text{(by (2))}. \end{aligned}$$

This proves the statement.

Proof of 10.12. Let T be an operator from ℓ_∞^m to ℓ_1^n. Recall that $\tilde{\pi}_2(T) \leq \pi_2(T)$. By 11.3 and 11.1, we have

$$\begin{aligned} \tilde{\pi}_2(T)^2 &\leq 2\sqrt{3}\, \pi_4(T)^2 \\ &\leq 2\sqrt{3}\, \pi_2(T)\|T\| \\ &\leq 2\sqrt{3}\, \tilde{\pi}_2(T)\|T\| , \end{aligned}$$

so $\tilde{\pi}_2(T) \leq 2\sqrt{3}\, \|T\|$. This is the required statement (with $K_G \leq 2\sqrt{3}$) .

Cotype 2 spaces are 2-dominated

A slight adaptation of the above proof gives the result stated. The following is a modified version of Lemma 11.3, in which the notion of cotype 2 effectively replaces Khinchin's inequality.

<u>11.4 Lemma.</u> Let $\kappa_2(X) = K$, and let T be an operator from ℓ_∞^m to X. Then $\pi_2(T) \leqslant 3^{1/4}K\, \pi_4(T)$.

Proof. With the notation of 11.3, we have $Ty_\varepsilon = \Sigma\, \varepsilon_i(Tx_i)$, and hence, by the definition of cotype 2,

$$
\begin{aligned}
(\Sigma\, \|Tx_i\|^2)^{\frac{1}{2}} &\leqslant K\Big(\frac{1}{2^k} \sum_\varepsilon \|Ty_\varepsilon\|^2\Big)^{\frac{1}{2}} \\
&\leqslant K\Big(\frac{1}{2^k} \sum_\varepsilon \|Ty_\varepsilon\|^4\Big)^{1/4} \\
&\leqslant K\, \pi_4(T)\, \Big\| \frac{1}{2^k}\, \Sigma y_\varepsilon^4 \Big\|^{1/4} \\
&\leqslant K\, \pi_4(T)\, 3^{1/4}\, \|\Sigma\, x_i^2\|^{\frac{1}{2}} \qquad \text{(by (2))}.
\end{aligned}
$$

<u>11.5 Theorem.</u> Every space of cotype 2 is 2-dominated. In fact, $\Delta_2(X) \leqslant \sqrt{3}\, \kappa_2(X)^2$.

Proof. Let $\kappa_2(X) = K$, and let T be an operator from ℓ_∞^m to X. By 11.4 and 11.1,

$$
\begin{aligned}
\pi_2(T)^2 &\leqslant \sqrt{3}\, K^2\, \pi_4(T)^2 \\
&\leqslant \sqrt{3}\, K^2\, \pi_2(T)\|T\|,
\end{aligned}
$$

so $\pi_2(T) \leqslant \sqrt{2}\, K^2\, \|T\|$.

Clearly, this theorem can be regarded as a generalization of the essence of Grothendieck's inequality. It was first proved by Maurey (1974a). Another proof was given by Rosenthal (1976). The method reproduced here - which is substantially shorter than the known alternatives - is due to Pisier (1978): the idea of it can be traced to Krivine (1973-74). Pisier in fact obtained both 11.5 and Grothendieck's inequality in a generalized form : the spaces ℓ_∞^m are replaced in the statements by C*-algebras (not necessarily commutative). His results have been further generalized by Haagerup (1985).

Using Khinchin's inequality for $r > 2$, one can modify 11.4 to obtain $\pi_2(T) \leq B_r\ \kappa_2(X)\ \pi_r(T)$. This yields the following variant of 11.5: $\Delta_2(X) \leq C_p\ \kappa_2(X)^p$ for any $p > 1$. However, the constant C_p tends to infinity as p tends to 1, and it is not known whether a similar statement holds with $p = 1$. Nor is it known whether, conversely, all 2-dominated spaces (or even all spaces with the Orlicz property) are of cotype 2. A conditional result in this direction will be described in section 12.

We outline briefly one alternative approach, also due to Pisier (for the details, see [FLO], chapter 4). For elements x_i of X, write

$$\rho_2^*(x_1, \dots ,x_n) \;=\; \sup\ \{\ \Sigma\, f_i(x_i) : f_i \in X^*,\ \rho_2(f_1, \dots ,f_n) \leq 1\ \}.$$

One shows that for $0 < \delta < 1$ and any 2-summing operator T,

$$\rho_2(Tx_1, \dots ,Tx_n) \;\leq\; [\ \delta\pi_2(T) + \delta^{-1/2}\|T\|\ ]\ \rho_2^*(x_1, \dots ,x_n) \qquad (1).$$

For x_i in ℓ_∞^m, it is easily shown from Khinchin's inequality that $\rho_2^*(x_1, \dots ,x_n) \leq \sqrt{2}\ \mu_2(x_1, \dots x_n)$. If X is of cotype 2 and T is an operator from ℓ_∞^m to X, it follows that

$$\pi_2(T) \;\leq\; \sqrt{2}\ \kappa_2(X)\ [\ \delta\pi_2(T) + \delta^{-1/2}\|T\|\ ]\ .$$

Write $\sqrt{2}\ \kappa_2(X) = K$, and put $\delta = 1/(2K)$. We obtain $\pi_2(T) \leq (2K)^{3/2}\|T\|$, so that $\Delta_2(X) \leq C\ \kappa_2(X)^{3/2}$ for a certain C. A refinement of the method shows that in fact $\Delta_2(X) \leq C'\kappa_2(X)\ \log(2\kappa_2(X))$.

The above approach is closely connected with the notion of factorization through a Hilbert space (and is presented in [FLO] in such terms). Given T in L(X,Y), let $\gamma_2(T)$ be the infimum of $\|T_2\|.\|T_1\|$ taken over all possible factorizations $T = T_2T_1$, where T_1 is in L(X,H) for some Hilbert space H, and T_2 is in L(H,Y). Of course, Pietsch's theorem shows that $\gamma_2(T) \leq \pi_2(T)$. A basic result on factorization is the theorem of Kwapien (1972), which states that $\gamma_2(TS) \leq \kappa_2(T)\tau_2(S)$ for any operators S,T. (Note that when applied to the identity in a space X, this shows that if X is both of type 2 and cotype 2, then X is isomorphic to a Hilbert space H, and $d(X,H) \leq \kappa_2(X)\,\tau_2(X)$). Now statement (1) applies in fact with π_2 replaced by γ_2. In combination with Kwapien's theorem, this leads to the following result, which Pisier calls the "abstract version of Grothendieck's theorem": if X* and Y are of cotype 2 and T is in FL(X,Y) (or is "approximable"), then $\gamma_2(T) \leq C\|T\|$, where $C = [2\ \kappa_2(X^*)\ \kappa_2(Y)]^{3/2}$.

12. RESULTS CONNECTED WITH THE BASIS CONSTANT

The basis constant

Let X be an n-dimensional normed linear space. Let $(b_1, \dots, b_n)$ be a basis of X, and let $(f_1, \dots, f_n)$ be the dual basis of X^*, so that $x = \sum_i f_i(x) b_i$ for all x in X. The "constant" (or "unconditional constant") of this basis, denoted by $\beta(b_1, \dots, b_n)$ is defined to be

$$\sup \{ \| \sum_i \lambda_i f_i(x) b_i \| : \|x\| \leq 1 \text{ and } |\lambda_i| \leq 1 \text{ for each } i \}.$$

Two equivalent ways of describing this quantity are, firstly,

$$\sup \{ \|T(\lambda_1, \dots, \lambda_n)\| : |\lambda_i| \leq 1 \text{ for each } i \},$$

where $T(\lambda_1, \dots, \lambda_n)$ is the operator that maps b_i to $\lambda_i b_i$ for each i, and secondly,

$$\sup \{ \mu_1[f_1(x) b_1, \dots, f_n(x) b_n] : \|x\| \leq 1 \}.$$

From the last expression, it is clear that we can restrict to $|\lambda_i| = 1$ in the previous ones.

The (unconditional) basis constant of the space X is defined to be the infimum of the constants of all bases of X.

First, some immediate remarks on these definitions :

(1) The usual basis of ℓ_p^n (for each p) has constant 1.

(2) For non-zero α_i, we have $\beta(\alpha_1 b_1, \dots, \alpha_n b_n) = \beta(b_1, \dots, b_n)$.

(3) In the real case, there is a lattice ordering associated with a basis: we define $x \leq y$ to mean $f_i(x) \leq f_i(y)$ for all i. Then $|x|$ is the element $\Sigma |f_i(x)| b_i$ (this notation makes sense in the complex case as well). The constant of the basis equates to $\sup \{ \|y\| : |y| \leq |x|, \|x\| \leq 1 \}$.

(4) The definition applies equally to an unconditional basis of an

infinite-dimensional Banach space. It is a standard consequence of the uniform boundedness theorem that the constant of such a basis is finite. Two such bases are said to be "equivalent" if there is an isomorphism mapping one onto the other.

12.1. A basis and its dual basis have the same constant. Hence $\beta(X^*) = \beta(X)$ for any finite-dimensional space X.

Proof. Let T be the operator such that $Tb_i = \lambda_i b_i$ for each i. It is easily verified that $T^* f_i = \lambda_i f_i$ for each i. The result follows, by the first equivalent form of the definition given above.

12.2. If X,Y have the same (finite) dimension, then $\beta(Y) \leq d(X,Y)\beta(X)$. Hence $\beta(X) \leq \sqrt{n}$ for any n-dimensional space X.

Proof. Let A be an isomorphism of X onto Y, and let $(b_1, \dots, b_n)$ be a basis of X, with constant β_0. Let T be the operator such that $Tb_i = \lambda_i b_i$ for each i. Then ATA^{-1} maps Ab_i to $\lambda_i(Ab_i)$. It follows that $\beta(Ab_1, \dots, Ab_n) \leq \beta_0 \|A\|.\|A^{-1}\|$.

The second statement follows, by 5.6.

Exercise. Let $(b_1, \dots, b_n)$ be a normalized basis of X, with constant β_0. Show that

$$d(X, \ell_\infty^n) \leq \beta_0^2 \, \|b_1 + \dots + b_n\| .$$

Note. Among infinite-dimensional spaces, only separable ones are candidates to have a basis. The following notion (defined on the pattern of $\mathcal{L}_\infty$-spaces) is not restricted in this way. The space X is said to have "local unconditional structure" with constant K if each finite-dimensional subspace E is contained in a larger finite-dimensional subspace F with $\beta(F) \leq K$. The infimum of such K is then called the "local unconditional structure constant" of X. It is known that all Banach lattices have local unconditional structure.

Connections with π_1 and the Gordon-Lewis constant

A number of results relate the basis constant to $d(X, \ell_1^n)$. The simplest one is:

<u>12.3.</u> If $\dim X = n$, then $d(X, \ell_1^n) \leqslant \beta(X)\, \pi_1^{(n)}(X)$.

Proof. Let $(b_1, \dots, b_n)$ be a normalized basis of X, with dual basis $(f_1, \dots, f_n)$ and constant β_0 . Since $x = \Sigma\, f_i(x) b_i$, we have $\|x\| \leqslant \Sigma\, |f_i(x)|$ for each x. If we show that for some C, we have $\Sigma\, |f_i(x)| \leqslant C\|x\|$ for all x, it follows that $d(X, \ell_1^n) \leqslant C$. Now by definition,

$$\mu_1[f_1(x)b_1, \dots, f_n(x)b_n] \leqslant \beta_0\|x\| . \tag{1}$$

Hence

$$\sum_1 |f_i(x)| = \sum_1 \|f_i(x)b_i\| \leqslant \beta_0 \pi_1^{(n)}(X)\, \|x\| .$$

The statement follows.

This is of very limited use, since $\pi_1(X)$ is at least $\sqrt{n}$. We now describe a more useful variant. For an operator T between finite-dimensional spaces, write

$$\gamma_1(T) = \nu_\infty(T^*) = \inf\, \{\mu_1(g_1, \dots, g_n) : T = \Sigma\, g_i \otimes y_i \text{ with each } \|y_i\| = 1\} .$$

Recall from 4.11 that this describes "factorization through ℓ_1".

<u>12.4.</u> For any operator T defined on X, $\gamma_1(T) \leqslant \beta(X)\, \pi_1(T)$.

Proof. Let $(b_1, \dots, b_n)$ be a normalized basis of X, with dual basis $(f_1, \dots, f_n)$ and constant β_0. Then

$$T = \sum_1 f_i \otimes (Tb_i) = \sum_1 g_i \otimes y_i ,$$

where $g_i = \|Tb_i\| f_i$ and $\|y_i\| = 1$. From (1),

$$\sum_1 \|f_i(x)(Tb_i)\| = \sum_1 |g_i(x)| \leqslant \beta_0\, \pi_1(T)\, \|x\| .$$

Hence $\mu_1(g_1, \dots, g_n) \leqslant \beta_0\, \pi_1(T)$, and the statement follows.

This result was formulated by Gordon & Lewis (1974), who used it to show that certain spaces have "large" basis constants. We give an account of this below. The quantity

$$\sup\, \{\gamma_1(T) : T \in L(X,Y),\ \pi_1(T) \leqslant 1\}$$

is consequently known as the <u>Gordon-Lewis constant</u> of the space X. Clearly, is not greater than $\beta(X)$.

The 2-dominated property and cotype 2

We saw in Section 11 that all spaces of cotype 2 are 2-dominated. With the help of the basis constant, we can prove a restricted converse. The method is essentially an abstract version of the proof that ℓ_1 is of cotype 2 (see Section 7). (The details are given for the real case; the complex case is the same with some modulus signs inserted).

Given a basis $(b_1,\dots,b_n)$, with dual basis $(f_1,\dots,f_n)$, we have already mentioned how the element $|x|$ can be defined. Similarly, it is natural to define multiplication "coordinatewise", so that, in particular, $x^2 = \sum_j f_j(x)^2 b_j$. Given elements $x_1,\dots,x_k$ of X, the element $z = (\sum_i x_i^2)^{1/2}$ can be defined in the same way. The 2-dominated constant $\Delta_2(X)$ now enables us to formulate an abstract analogue of 7.5.

<u>12.5</u> Let $\beta(b_1,\dots,b_n) = \beta_0$, and let $z = (\sum_i x_i^2)^{1/2}$ in the sense just indicated. Then

$$(\Sigma \|x_i\|^2)^{1/2} \leqslant \beta_0 \Delta_2(X) \|z\| .$$

Proof. We have $f_j(z)^2 = \sum_i f_j(x_i)^2$ for each j. Define an operator T from ℓ_∞^n to X by :

$$Ty = \sum_j y(j) f_j(z) b_j .$$

By the definition of the basis constant, if $\|y\|_\infty \leqslant 1$, then $\|Ty\| \leqslant \beta_0 \|z\|$. Hence $\|T\| \leqslant \beta_0 \|z\|$, so $\pi_2(T) \leqslant \beta_0 \Delta_2(X) \|z\|$.

We obtain elements y_i of ℓ_∞^n such that $Ty_i = x_i$ by choosing $y_i(j)$ so that $y_i(j) f_j(z) = f_j(x_i)$ (if $f_j(z) = 0$, put $y_i(j) = 0$). Then $\sum_i y_i(j)^2 \leqslant 1$ for all j, so $\mu_2(y_1,\dots,y_k) \leqslant 1$. Hence

$$(\Sigma \|x_i\|^2)^{1/2} \leqslant \pi_2(T) \leqslant \beta_0 \Delta_2(X) \|z\| .$$

<u>12.6 Lemma.</u> Let the operation | | be defined relative to a basis (with constant β_0) as above. If $|u| \leqslant \sum_i |x_i|$, then $\|u\| \leqslant \beta_0 \sum_i \|x_i\|$.

Proof. For each j, we have $|f_j(u)| \leqslant \sum_i |f_j(x_i)|$. Hence we can choose α_{ij} with $|\alpha_{ij}| \leqslant 1$ and

$$f_j(u) = \sum_i \alpha_{ij} f_j(x_i)$$

for each j. Define elements y_i by setting

$$f_j(y_i) = \alpha_{ij} f_j(x_i) .$$

Then $u = \sum_i y_i$ and $\|y_i\| \leq \beta_0 \|x_i\|$ for each i.

12.7 Proposition. For any finite-dimensional space X, we have

$$\kappa_2(X) \leq \sqrt{2}\, \beta(X)^2\, \Delta_2(X) .$$

Proof. Let $(b_1, \dots ,b_n)$ be a basis with constant β_0, and let $|\ |$ and multiplication be defined relative to this basis. Choose elements $x_1, \dots ,x_k$ of X, and let $z = (\Sigma\, x_i^2)^{1/2}$. By Khinchin's inequality,

$$z \leq \frac{\sqrt{2}}{2^k} \sum_{\varepsilon \in D_k} |\sum_i \varepsilon_i x_i| .$$

Hence, by 12.6,

$$\|z\| \leq \frac{\sqrt{2}}{2^k}\, \beta_0 \sum_{\varepsilon} \| \sum_i \varepsilon_i x_i\| = \sqrt{2}\, \beta_0\, \rho_1(x_1, \dots ,x_k) .$$

The inequality in 12.5 now gives the result.

Clearly, the same applies to infinite-dimensional spaces with the local unconditional structure constant replacing $\beta(X)$.

Connections with Grothendieck's inequality

The next group of results originates with Lindenstrauss & Pelczynski (1968). We start with the promised "converse" of Grothendieck's inequality.

12.8 Proposition. (i) Let X be an n-dimensional space. Suppose that X has the Grothendieck property with constant K, and that $\pi_{2,1}(X) = A$ (note that $A \leq K$). Then $d(X, \ell_1^n) \leq A K \beta(X)^2$.

(ii) Let X be an infinite-dimensional space with unconditional basis, and let A,K be as in (i). Then $d(X, \ell_1) \leq A K \beta(X)^2$.

Proof. For (i), let $(b_1, \dots ,b_n)$ be a normalized basis with constant β_0, and let $(f_1, \dots ,f_n)$ be the dual basis. For all x in X, we have

$$\mu_1[f_1(x)b_1, \dots ,f_n(x)b_n] \leq \beta_0 \|x\| \qquad (1),$$

and therefore

$$\sum_i f_i(x)^2 \leqslant A^2\beta_0^2 \|x\|^2 \qquad (2).$$

Let T be the operator from X to ℓ_2^n defined by $Tb_i = e_i$, so that $\|Tx\|^2 = \sum_i f_i(x)^2$. By (2), $\|T\| \leqslant A\beta_0$, so we have $\pi_1(T) \leqslant AK\beta_0$. From (1), it now follows that

$$\|x\| \leqslant \sum_i |f_i(x)| = \sum_i \|T(f_i(x)b_i)\| \leqslant AK\beta_0^2 \|x\| .$$

The statement follows, and the proof of (ii) is exactly similar.

12.9 Corollary. Let X be a complemented subspace of an $\mathcal{L}_1$-space (with projection P). Suppose that X has an unconditional basis. Then X is isomorphic to ℓ_1 (or ℓ_1^n), with

$$d(X, \ell_1) \leqslant \sqrt{2}\, K_G\, \beta(X)^2 \|P\| .$$

Proof. The space X has the Grothendieck property with constant $K_G\|P\|$ (see 10.17), and $\pi_{2,1}(X) \leqslant \sqrt{2}$.

12.10 Corollary. Every unconditional basis of ℓ_1 is equivalent to the usual basis.

Proof. The isomorphism constructed in 12.8 maps b_i to e_i.

We observed in Section 4 that it is not known whether there is a constant C such that $d(X, \ell_\infty^n) \leqslant C\lambda(X)$ for all n-dimensional spaces X. By applying 12.9 to the dual space, we can obtain a partial result in this direction.

12.11 Proposition. For any n-dimensional space X,

$$d(X, \ell_\infty^n) \leqslant \sqrt{2}\, K_G\, \beta(X)^2 \lambda(X)^2 .$$

Proof. We may assume that X is a subspace of some ℓ_∞^N. Let P be a projection of ℓ_∞^N onto X with $\|P\|$ close to $\lambda(X)$. Then P^* maps X^* into ℓ_1^N, and if R is the "restriction" operator from ℓ_1^N to X^*, then $RP^* = I_{X^*}$, since for $f \in X^*$ and $x \in X$, we have $(P^*f)(x) = f(Px) = f(x)$. Hence $\|P^*f\| \geqslant \|f\|$ for $f \in X^*$, so $d[X^*, P^*(X^*)] \leqslant \|P\|$.

It also follows that P^*R is a projection of ℓ_1^N onto $P^*(X^*)$. Since $\beta(X^*) = \beta(X)$ and $\|P^*R\| \leqslant \|P\|$, we have from 12.9 :

$$d[P^*(X^*), \ell_1^n] \leqslant \sqrt{2}\, K_G\, \beta(X)^2 \|P\| .$$

Hence

$$d(X,\ell_\infty^n) = d(X^*,\ell_1^n) \leq \sqrt{2}\, K_G\, \beta(X)^2\, \|P\|^2 .$$

There is also a dual version of 12.10:

12.12. Every unconditional basis of c_0 is equivalent to the usual basis.

Proof. (For this, we assume familiarity with some elementary facts on unconditional bases). Let (b_n) be such a basis (normalized). The dual basic sequence (f_n) is unconditional and bounded. Applying 12.9 to each $(f_1, \ldots ,f_n)$, we see that (f_n) is equivalent to the usual basis of ℓ_1. It follows by standard arguments that (b_n) is equivalent to the usual basis of c_0.

Spaces with large basis constants

The problem of finding finite-dimensional spaces with arbitrarily large basis constants was resolved very satisfactorily by Gordon & Lewis (1974), who showed that the "natural" space of operators $L(\ell_2^n)$ has basis constant not less than $\frac{1}{4}\sqrt{n}$. We present here a simplified version of their proof due to Schütt (1978). As already mentioned, the method depends on 12.4, and consequently can be regarded as another application of the theory of summing and nuclear norms. It is an ingenious and elegant piece of work.

Let L_n be the space $L(\mathbb{R}^n)$, identified with $n \times n$ matrices in the usual way : $Ae_j = \sum_i a_{ij}\, e_i$. Regarding L_n as simply $\mathbb{R}^{n^2}$, we write

$$\langle A,B \rangle = \sum_i \sum_j a_{ij} b_{ij}$$

and $\|A\|_2 = \langle A,A \rangle^{1/2}$. Given any norm α on L_n, let α^* be the dual norm on L_n, identified with its own dual in this way (there is no need to think in terms of trace duality !).

Recall than D_n denotes the set of elements $\delta = (\delta_1, \ldots ,\delta_n)$ with each δ_i in $\{-1,1\}$. Given $\delta \in D_n$, let U_δ be the corresponding diagonal operator on $\mathbb{R}^n$. Then $U_\varepsilon A U_\delta$ is the operator with matrix $(\delta_i\, a_{ij}\, \varepsilon_j)$. We shall say that a norm α on L_n is <u>invariant</u> if $\alpha(U_\varepsilon A U_\delta) = \alpha(A)$ for all A,δ,ε. Since $\langle U_\varepsilon A U_\delta ,B\rangle = \langle A,U_\varepsilon B U_\delta \rangle$, it follows that α^* is then invariant. For any p,q, the ordinary operator norm of $L(\ell_p^n,\ell_q^n)$ is invariant, since U_δ,U_ε are isometries.

We start with a bivariate version of Khinchin's inequality:

12.13 With the above notation,

$$\|A\|_2 \leq \frac{1}{2^{2n-1}} \sum_{\delta \in D_n} \sum_{\varepsilon \in D_n} \left| \sum_i \sum_j \delta_i a_{ij} \varepsilon_j \right| .$$

Proof. Define elements a_i of ℓ_2^n by : $a_i(j) = a_{ij}$. Then $\|A\|_2^2 = \sum_i \|a_i\|_2^2$. By 7.4 (Khinchin's inequality for Hilbert spaces).

$$\|A\|_2 = \left(\sum_i \|a_i\|_2^2 \right)^{1/2} \leq \frac{\sqrt{2}}{2^n} \sum_{\delta \in D_n} \left\| \sum_i \delta_i a_i \right\|_2 .$$

By the ordinary form of Khinchin's inequality,

$$\left\| \sum_i \delta_i a_i \right\|_2 \leq \frac{\sqrt{2}}{2^n} \sum_{\varepsilon \in D_n} \left| \langle \varepsilon, \sum_i \delta_i a_i \rangle \right|$$

$$= \frac{\sqrt{2}}{2^n} \sum_{\varepsilon \in D_n} \left| \sum_i \sum_j \delta_i a_{ij} \varepsilon_j \right| .$$

Given elements A,B of L_n, write A.B for the "pointwise" product $(a_{ij}b_{ij})$, and let M_B be the "multiplication" operator on L_n defined by : $M_B(A) = A.B$.

12.14. Let $S = (s_{ij})$ and $\Delta = (\delta_{ij})$ be elements of L_n, with $\delta_{ij} \in \{-1,1\}$ for each i,j. Let α be an invariant norm on L_n, and regard $M_{\Delta.S}$ as an operator from (L_n,α) to $(L_n, \| \ \|_2)$. Then $\pi_1(M_{\Delta.S}) \leq 2\alpha^*(S)$.

Proof. By 12.13, we have

$$\|M_{\Delta.S}(A)\|_2 = \|S.A\|_2$$

$$\leq \frac{1}{2^{2n-1}} \sum_\delta \sum_\varepsilon \left| \langle A, U_\varepsilon S U_\delta \rangle \right| .$$

As a functional on (L_n,α), the norm of $U_\varepsilon S U_\delta$ is $\alpha^*(S)$. There are 2^{2n} such functionals. The statement follows, by 3.2.

12.15 Example. Let J_n, K_n be the formal identity operators from $[L(\ell_\infty^n, \ell_1^n), \| \ \|]$ and $[L(\ell_2^n), \| \ \|]$ (respectively) into $(L_n, \| \ \|_2)$. In 12.14, take $s_{ij} = \delta_{ij} = 1$ for all i,j. Then $\langle A,S \rangle = \sum_i \sum_j a_{ij} = \langle Ae,e \rangle$, where $e = (1, \ldots ,1)$. From this, we see that $\alpha^*(S)$ is equal to 1 and n in the two cases. Hence we have $\pi_1(J_n) \leq 2$ and $\pi_1(K_n) \leq 2n$.

<u>Exercise.</u> Show that $\pi_2(J_n) = 1$ and $\pi_2(K_n) = n$.

<u>12.16.</u> Let $T = (t_{ij})$ and $\Delta = (\delta_{ij})$ be elements of L_n, with $\delta_{ij} \in \{-1,1\}$ for each i,j. Let α be an invariant norm on L_n, and regard M_T as an operator from $(L_n, \| \ \|_2)$ to (L_n,α). Then $\nu_\infty(M_T) \geq \frac{1}{2}\alpha(\Delta.T)$.

Proof. We use trace duality. Let E_{ij} be the matrix having 1 in place (i,j) and 0 elsewhere. Such matrices form the "natural" basis of L_n. Clearly, $M_A(E_{ij}) = a_{ij}E_{ij}$, and hence we have $\text{trace}\,(M_A M_B) = \langle A,B \rangle$ for any A,B in L_n.

There is an element S of L_n such that $\alpha^*(S) = 1$ and $\langle S, \Delta.T \rangle = \alpha(\Delta.T)$. But

$$\begin{aligned}\langle S, \Delta.T \rangle &= \langle \Delta.S, T \rangle \\ &= \text{trace}\,(M_{\Delta.S}M_T) \\ &\leq \pi_1(M_{\Delta.S})\ \nu_\infty(M_T)\end{aligned}$$

where $M_{\Delta.S}$ is regarded as an operator from (L_n,α) to $(L_n, \| \ \|_2)$. By 12.14, $\pi_1(M_{\Delta.S}) \leq 2$, so the statement follows.

<u>12.17.</u> Let α be an invariant norm on L_n. Suppose that for a certain S, Δ in L_n we have $\delta_{ij} \in \{-1,1\}$ for all i,j and $\alpha(\Delta.S) = K\alpha(S)$. Then $\beta(L_n,\alpha) \geq \frac{1}{4}K$.

Proof. We show in fact that $\beta(L_n,\alpha^*) \geq \frac{1}{4}K$; the result then follows, by 12.1. Regard M_S as an operator from (L_n,α^*) to $(L_n, \| \ \|_2)$. By 12.14, $\pi_1(M_S) \leq 2\alpha(S)$.

From the fact that $M_S(E_{ij}) = s_{ij}E_{ij}$, it is easily seen that the dual operator $M_S{}^*$ identifies with M_S, regarded now as an operator from $(L_n, \| \ \|_2)$ to (L_n,α) (note that M_S is just a diagonal operator with respect to this basis). By 12.16, $\nu_\infty(M_S{}^*) \geq \frac{1}{2}\alpha(\Delta.S) \geq \frac{1}{2}K\alpha(S)$. But $\nu_\infty(M_S{}^*) = \gamma_1(M_S)$, so the statement now follows from 12.4.

Note that the hypothesis of 12.17 amounts to saying that the natural basis (L_n,α) has constant at least K. We now use the "Littlewood matrices" to show that for certain choices of α, this is indeed the case (with $K = \sqrt{n}$). These matrices are defined inductively as follows:

$$W_1 = \begin{bmatrix} 1 & 1 \\ 1 & -1 \end{bmatrix}, \qquad W_k = \begin{bmatrix} W_{k-1} & W_{k-1} \\ W_{k-1} & -W_{k-1} \end{bmatrix}.$$

W_k is a symmetric $n \times n$ matrix (where $n = 2^k$) with orthogonal rows, so that $W_k{}^2 = nI$, and

$$\|W_k x\|_2 = \sqrt{n}\ \|x\|_2 \tag{1}$$

for all x in $\mathbb{R}^n$. The entries are all 1 or -1.

<u>12.18 Theorem.</u> For $n = 2^k$, each of the following spaces (with operator norm) has basis constant not less than $\frac{1}{4}\sqrt{n}$:

$$L(\ell_2^n),\quad L(\ell_\infty^n, \ell_1^n),\quad L(\ell_\infty^n, \ell_2^n)\ .$$

Proof. Let E be the matrix with all entries equal to 1. This corresponds to the operator $e \otimes e$. In each case, we compare $\|E\|$ with $\|W_k\|$, and apply 12.17.

(i) $L(\ell_2^n)$: here $\|E\| = \|e\|_2^2 = n$, while $\|W_k\| = \sqrt{n}$, by (1).

(ii) $L(\ell_\infty^n, \ell_1^n)$: now $\|E\| = \|e\|_1{}^2 = n^2$, and by (1),

$$\|W_k x\|_1 \leq \sqrt{n}\ \|W_k x\|_2 = n\|x\|_2 \leq n\sqrt{n}\ \|x\|_\infty ,$$

so $\|W_k\| \leq n\sqrt{n}$.

(iii) $L(\ell_\infty^n, \ell_2^n)$: now $\|E\| = \|e\|_1\|e\|_2 = n\sqrt{n}$, while $\|W_k\| \leq n$.

<u>Notes.</u> (1) The tensor product notation for these three spaces is (respectively) : $\ell_2^n \otimes \ell_2^n$, $\ell_1^n \otimes \ell_1^n$, $\ell_1^n \otimes \ell_2^n$. Of course, the third space equates with $L(\ell_2^n, \ell_1^n)$ by conjugacy.

(2) For T in $L(\ell_2^n)$, we have $\|T\| \leq \|T\|_2 \leq \sqrt{n}\ \|T\|$ (note that $\|T\|_2 = \pi_2(T)$). Hence the basis constant is not greater than $\sqrt{n}$.

<u>Exercise.</u> Show that in $L(\ell_1^n)$ the natural basis has constant 1. Consider the remaining cases of $L(\ell_p^n, \ell_q^n)$ (where p,q are 1,2 or ∞).

<u>Exercise.</u> Use the Littlewood matrices to show that for $n = 2^k$, $d(\ell_1^n, \ell_\infty^n) \leq \sqrt{n}$.

Of course, the dimension of L_n is n^2. It has in fact been shown (by a specially constructed example, not a "natural" one) that there exist C and n-dimensional spaces X_n for each n such that $\beta(X_n) \geq C\sqrt{n}$. See Pisier [FLO], 8e.

13. ESTIMATION OF SUMMING NORMS USING A RESTRICTED NUMBER OF ELEMENTS

Elementary facts

Recall that $\pi_p^{(n)}(T)$ is defined like $\pi_p(T)$, but considering only finite sequences of length not more than n. We have already mentioned a number of sporadic elementary results about $\pi_p^{(n)}$. For instance, for T in $L(\ell_2^m, \ell_2^n)$, we have $\pi_2^{(m)}(T) = \pi_2(T)$ (3.9), and for all operators into ℓ_∞^n, we have $\pi_p^{(n)}(T) = \pi_p(T)$ (5.5). (See also 3.14, 6.9, 7.18).

In this section, we will prove two results relating $\pi_2^{(n)}(T)$ to $\pi_2(T)$ for arbitrary finite-rank operators T. In both cases we make use of the following simple lemma.

<u>13.1 Lemma.</u> Let Y be a normed linear space, n a positive integer. Suppose that for some m,K, we have $\pi_2(A) \leqslant K\pi_2^{(m)}(A)$ for all A in $L(\ell_2^n, Y)$. Then for all operators mapping into Y (from any normed linear space) with rank n, we have $\pi_2(T) \leqslant K\pi_2^{(m)}(T)$.

Proof. Take $\varepsilon>0$. By 3.6, there is an operator A in $L(\ell_2^n, Y)$ with $\|A\| = 1$ and $\pi_2(TA) \geqslant (1-\varepsilon)\pi_2(T)$. By hypothesis,

$$\pi_2(TA) \leqslant K\pi_2^{(m)}(TA) \leqslant K\pi_2^{(m)}(T).$$

The statement follows.

As a first consequence of this, we have at once:

<u>13.2.</u> For all operators mapping into ℓ_2^n, we have $\pi_2(T) = \pi_2^{(n)}(T)$.

Proof. As mentioned above, $\pi_2(A) = \pi_2^{(n)}(A)$ for all elements A of $L(\ell_2^n)$.

For operators on a finite-dimensional space X, we remark that $\pi_p^{(m)}(T)$ is actually attained at a certain finite sequence $(a_1, \dots, a_m)$, since the set of all finite sequences with $\mu_p(x_1, \dots, x_m) \leq 1$ is clearly a compact set in the space X^m.

Estimation of π_2 using n elements

The following theorem of Tomczak-Jaegermann (1979) gives a very satisfactory solution to the problem for π_2.

13.3 Theorem. For any operator T of rank n, we have $\pi_2^{(n)}(T) \geq \frac{1}{2}\pi_2(T)$.

Proof. By 13.1, it is sufficient to prove the statement for operators in T in L(H,Y), where $H = \ell_2^n$. Assume that $\pi_2(T) = 1$. By Pietsch's theorem, T = WV, where V is in $L(H,H_1)$, W is in $L(H_1,Y)$, $\pi_2(V) = \pi_2(T)$, $\|V\| = 1$ and H_1 is another Hilbert space. We will show that there is an orthonormal basis $(a_1, \dots, a_n)$ of H such that $\Sigma \|Ta_i\|^2 \geq \frac{1}{4}$: it then follows that $\pi_2^{(n)}(T) \geq \frac{1}{2}$.

Define the a_i as follows. Since $\|W\| = 1$, we can choose a_1 such that $\|a_1\| = 1$ and

$$\|Ta_1\| = \|WVa_1\| = \|Va_1\| .$$

Having chosen $a_1, \dots, a_{j-1}$, let $E_j = \{a_1, \dots, a_{j-1}\}^{\perp}$ and let W_j be the restriction of W to $V(E_j)$. Choose a_j in E_j such that $\|a_j\| = 1$ and

$$\|Ta_j\| = \|WVa_j\| = \|W_j\| \cdot \|Va_j\| .$$

Note that

$$\pi_2(V)^2 = \sum_{j=1}^{n} \|Va_j\|^2 = 1 .$$

Clearly, $\|W_j\| \geq \|W_{j+1}\|$ for each j. Let m be the largest integer such that $\|W_m\| \geq 1/\sqrt{2}$. and write

$$\sum_{j=1}^{m} \|Va_j\|^2 = \alpha.$$

Then $\|Ta_j\|^2 \geq \frac{1}{2}\|Va_j\|^2$ for $j \leq m$, so

$$\sum_{j=1}^{m} \|Ta_j\|^2 \geq \frac{\alpha}{2} .$$

The proof will be complete if we can show that $\alpha \geq \frac{1}{2}$. The strategy is to express T as $T_1 + T_2$ and use the fact that $\pi_2(T_1) + \pi_2(T_2) \geq 1$.

Let P be the orthogonal projection onto E_{m+1}, so that $Pa_j = 0$ for $j \leq m$ and $Pa_j = a_j$ for $j > m$. Then

$$\pi_2(VP)^2 = \sum_{m+1}^{n} \|Va_j\|^2 = 1 - \alpha .$$

Choose $\delta > 0$ and let $T_2 = \delta WVP$, $T = T_1 + T_2$. Now $WVP = W_{m+1}VP$ and $\|W_{m+1}\| < 1\sqrt{2}$, so $\pi_2(T_2)^2 \leq \frac{1}{2}\delta^2(1-\alpha)$. Also, $T_1 = W(V - \delta VP)$. For $j > m$, $(V - \delta VP)a_j = (1-\delta)Va_j$, so

$$\pi_2(T_1)^2 \leq \pi_2(V - \delta VP)^2 = \alpha + (1-\delta)^2(1-\alpha)$$
$$= 1 - (2\delta - \delta^2)(1-\alpha).$$

Write $\pi_2(T_j) = r_j$ for $j = 1,2$. Then $r_1+r_2 \geq 1$ and $r_2 \leq \delta\sqrt{1-\alpha}/\sqrt{2}$, so

$$1 \leq (r_1 + r_2)^2 = 2r_1^2 + 2r_2^2 - (r_1 - r_2)^2$$

$$\leq 2 - 2(2\delta-\delta^2)(1-\alpha) + \delta^2(1-\alpha) - \left[1 - \delta\sqrt{2(1-\alpha)}\right]^2 .$$

On simplifying and dividing by δ, this inequality becomes

$$4(1-\alpha) \leq 2\sqrt{2(1-\alpha)} + \delta(1-\alpha).$$

This holds for all $\delta > 0$, and hence also with $\delta = 0$ to give $\sqrt{1-\alpha} \leq 1/\sqrt{2}$, so that $1-\alpha \leq \frac{1}{2}$, as required.

For the identity in an n-dimensional space, the factor $\frac{1}{2}$ can be improved at least to $1/\sqrt{3}$ (see [FDBS], lecture 17).

The next example shows that $\pi_2^{(n)}$ does not simply coincide with π_2, even when $n = 2$.

<u>13.4 Example.</u> Let H_2 be the subspace of ℓ_∞^3 consisting of elements with $\Sigma x(i) = 0$. Let a_1,a_2 be elements of H_2 with $\mu_2(a_1,a_2) = 1$, and let $\|a_1\|^2 + \|a_2\|^2 = 2-\varepsilon$ (where $\varepsilon \leq \frac{1}{4}$). We show that $\varepsilon \geq \frac{1}{32}$.

Note that $1-\varepsilon \leq \|a_i\| \leq 1$ for each i, and that $a_1(j)^2 + a_2(j)^2 \leq 1$ for each j. We may assume that $\|a_1\| = |a_1(1)|$ and $\|a_2\| = |a_2(2)|$. Then $|a_2(1)|, |a_1(2)| \leq \delta$, where $(1-\varepsilon)^2 + \delta^2 = 1$. By the defining property of H_2, it now follows that $|a_i(3)| \geq 1-\varepsilon-\delta$ for $i = 1,2$, and hence that $1-\varepsilon-\delta \leq 1/\sqrt{2}$. Write $c = 1-1/\sqrt{2}$. Then $\delta \geq c-\varepsilon$, so $\delta^2 = 2\varepsilon-\varepsilon^2 \geq (c-\varepsilon)^2$, hence $2\varepsilon \geq c^2 - 2c\varepsilon$, giving $\varepsilon > \frac{1}{32}$, as stated.

The number of elements required for exact determination of π_2

The next result first appeared in [FDBS], where it is attributed to T. Figiel. First, we require two simple algebraic lemmas.

13.5 Lemma. Let X be an n-dimensional real linear space, and let $x_1, \ldots, x_k$ be elements of X, where $k > n$. Then there exist non-negative scalars λ_i such that $\Sigma\, x_i = \Sigma\, \lambda_i x_i$ and $\lambda_i > 0$ for at most n values of i.

Proof. There are scalars μ_i, not all zero, such that $\Sigma\, \mu_i x_i = 0$. By taking a suitable scalar multiple, we can ensure that $\max \mu_i = 1$. Then $\Sigma\, x_i = \Sigma(1-\mu_i)x_i$. This is an expression with non-negative coefficients $1-\mu_i$, at least one of which is 0. Proceeding in this way, we can eliminate $k-n$ of the x_i.

13.6 Lemma. Let X be an n-dimensional real linear space, and let $N = \frac{1}{2}n(n+1)$. Let $x_1, \ldots, x_k$ be elements of X, where $k > N$. Then there exist scalars λ_i, non-zero for at most N values of i, such that for all f in X^*,

$$\sum_i f(x_i)^2 = \sum_i \lambda_i^2 f(x_i)^2 .$$

Proof. Let $B_s(X^*)$ be the space of symmetric bilinear forms on X^*. This corresponds to the space of symmetric $n\times n$ matrices, so has dimension N. For each i, define ϕ_i in $B_s(X^*)$ by : $\phi_i(f,g) = f(x_i)g(x_i)$. By 13.5, we can express $\Sigma\, \phi_i$ as $\Sigma\, \lambda_i^2\phi_i$ with at most N of the λ_i non-zero.

13.7 Proposition. Let T be an operator of rank n between real normed linear spaces. Then $\pi_2(T) = \pi_2^{(N)}(T)$, where $N = \frac{1}{2}n(n+1)$.

Proof. By 13.1, it is sufficient to prove the statement for an operator T defined on ℓ_2^n. We show that $\pi_2^{(k)}(T) = \pi_2^{(k-1)}(T)$ for all $k > N$.

Choose $x_1, \ldots, x_k$ in ℓ_2^n with $\mu_2(x_1, \ldots, x_k) = 1$. Let the numbers λ_i be as in 13.6, and let $\lambda_0 = \max \lambda_i$, $\mu_0 = 1/\lambda_0$ (clearly $\lambda_0 \geqslant 1$). Define

$$y_i = \mu_0 \lambda_i x_i, \quad z_i = (1-\mu_0^2\lambda_i^2)^{1/2}\, x_i.$$

Then $\|Tx_i\|^2 = \|Ty_i\|^2 + \|Tz_i\|^2$. At most N of the y_i are non-zero, and at most k-1 of the z_i (since $\lambda_0 = \lambda_i$ for some i). For any f in X^*, we have from 13.6:

$$\sum_i f(y_i)^2 = \mu_0^2 \sum_i \lambda_i^2 f(x_i)^2 = \mu_0^2 \sum_i f(x_i)^2 \leqslant \mu_0^2 ,$$

$$\sum_i f(z_i)^2 = \sum_i (1-\mu_0^2\lambda_i^2) f(x_i)^2 = (1-\mu_0^2) \sum_i f(x_i)^2 \leq 1-\mu_0^2 .$$

If $\pi_2^{(k-1)}(T) = M$, it follows that

$$\sum_i \|Ty_i\|^2 \leq \mu_0^2 M^2, \qquad \sum_i \|Tz_i\|^2 \leq (1-\mu_0^2)M^2,$$

so that $\sum_i \|Tx_i\|^2 \leq M^2$. Hence $\pi_2^{(k)}(T) = M$, as stated.

The same method applies in the complex case, but in 13.6 we must use hermitian forms on X^*. It is easily checked that the set of such forms is a real linear space of dimension n^2, so the conclusion holds with N equal to n^2.

Estimation of π_p for other p.

We have seen (5.5) that for operators mapping into ℓ_∞^N, we have $\pi_p(T) = \pi_p^{(N)}(T)$. It is possible to deduce a statement for general T with the help of the following variant of the basic embedding theorem 0.13, in which the dimension of the containing space ℓ_∞^N is limited.

13.8 Lemma. Let Y be an n-dimensional real normed linear space. Then there exist $N \leq 4^n$ and an operator A of Y into ℓ_∞^N such that $\frac{1}{3}\|y\| \leq \|Ay\| \leq \|y\|$ for all y in Y.

Proof. Let $f_1, \dots, f_N$ be a maximal set in U_{Y^*} such that $\|f_i - f_j\| > \frac{2}{3}$ for $i \neq j$. Given $y \in Y$, there exists f in U_{Y^*} such that $f(y) = \|y\|$. For some i, we have $\|f - f_i\| \leq \frac{2}{3}$ (otherwise f could be added to the set), and hence $|f_i(y)| \geq \frac{1}{3}\|y\|$. The operator A is defined by : $(Ay)(i) = f_i(y)$. It remains to estimate N.

Identifying Y^* with $\mathbb{R}^n$, let V be the "volume" of U_{Y^*} (i.e. ordinary Lebesgue measure). The balls $B(f_i, \frac{1}{3})$ are disjoint and have volume $V/3^n$. They are all contained in $\frac{4}{3}U_{Y^*}$, which has volume $4^n V/3^n$. It follows that $N \leq 4^n$.

13.9 Proposition. Let T be an operator of rank n between real normed linear spaces, and let $N = 4^n$. Then $\pi_p^{(N)}(T) \geq \frac{1}{3}\pi_p(T)$ for all p.

Proof. Let T be in L(X,Y). By 13.8, there is an operator A of T(X) into ℓ_∞^N such that $\|A\| = 1$ and $\|A^{-1}\| \leq 3$. By the result quoted above, $\pi_p(AT) = \pi_p^{(N)}(AT)$. Hence

$$\pi_p(T) \leq \|A^{-1}\|\pi_p(AT) \leq 3\pi_p^{(N)}(AT) \leq 3\pi_p^{(N)}(T).$$

The same applies in the complex case, with $N = 4^{2n}$.

An example is given in [FDBS], lecture 21 of operators T_n of rank n (for each n) for which $\pi_1(T) = \sqrt{n}$, while for a certain C (independent of n) $\pi_1^{(k)}(T) \leq C\sqrt{\log k}$ for each k.

14. PISIER'S THEOREM FOR $\pi_{2,1}$

In this section we present a restricted "Pietsch"-type theorem for the mixed summing norm $\pi_{2,1}$ (or $\pi_{q,p}$ with $q > p$), due to Pisier (1986). The restriction is that it applies to operators defined on the whole of an L_∞-space X, not on a subspace.

First, let us describe the easy implication. Suppose that there is a positive functional ϕ on X such that for some C, we have $\|Tx\|^2 \leq C^2\phi(|x|)\,\|x\|$ for all x. If we take elements x_i such that $\|\,\Sigma\,|x_i|\,\| \leq 1$, then certainly $\|x_i\| \leq 1$ for each i, so

$$\sum_i \|Tx_i\|^2 \leq C^2\,\phi(\,\sum_i |x_i|\,) \leq C^2 .$$

Hence $\pi_{2,1}(T) \leq C$.

Pisier's theorem states the converse, exept that (curiously enough) we do not obtain C exactly equal to $\pi_{2,1}(T)$. The proof is ingenious, and quite different from the proof of Pietsch's theorem. It uses both the multiplication in L_∞ and (in infinite dimensions) the weak-star compactness of the dual unit ball.

<u>14.1 Theorem.</u> Let X be any of ℓ_∞^n, $\ell_\infty(s)$, $L_\infty(\mu)$, C(K), and let T be an operator defined on X with $\pi_{2,1}(T)$ finite. Then there is a positive functional ϕ on X such that $\|\phi\| = 1$ and for all $x \in X$,

$$\|Tx\|^2 \leq 2\,\pi_{2,1}(T)^2\,\phi(|x|)\,\|x\| .$$

Proof. We may assume that $\pi_{2,1}(T) = 1$. Write e for the constant function 1 (in X). Fix n. There exist elements $x_1, \dots, x_{k(n)}$ such that $\Sigma|x_i| \leq e$ and $\Sigma\|Tx_i\|^2 = C_n^{\,2}$, where $C_n \geq 1 - \frac{1}{n}$. Take scalars α_i such that $\Sigma\alpha_i^{\,2} = 1$ and $\Sigma\alpha_i\|Tx_i\| = C_n$. There are functionals f_i in Y^* such that $\|f_i\| = \alpha_i$ and $f_i(Tx_i) = \alpha_i\|Tx_i\|$, hence $\Sigma f_i(Tx_i) = C_n$.

Define a functional ϕ_n on X by

$$\phi_n(x) = \sum_1 f_i[T(xx_i)] .$$

Clearly, $\phi_n(e) = C_n$. We show that $\|\phi_n\| \leq 1$. Let $\|x\| \leq 1$. Then $\Sigma |xx_i| \leq e$, so $\Sigma \|T(xx_i)\|^2 \leq 1$. By Schwarz's inequality, it follows that

$$|\phi_n(x)| \leq \sum_1 \alpha_i \|T(xx_i)\| \leq 1 .$$

If $C_n = 1$ for some n, let $\phi = \phi_n$. Otherwise, let ϕ be a weak-star cluster point of the sequence (ϕ_n) defined in this way (in finite dimensions, this just means the limit of a norm-convergent subsequence). Then $\phi(e) = 1$ and $\|\phi\| = 1$: this implies that ϕ is positive. It remains to prove the stated inequality.

Choose $y \in X$ with $\|y\| \leq 1$. Write $k(n) = k$, and let

$$y_i = x_i(e - |y|) \qquad \text{for } 1 \leq i \leq k,$$
$$y_{k+1} = y.$$

Then $\sum_1^{n+1} |y_i| \leq e - |y| + |y| = e$, so $\sum_1^{n+1} \|Ty_i\|^2 \leq 1$. Hence

$$\phi_n(e - |y|) = \sum_1^n f_i(Ty_i) \leq \left(\sum_1^n \|Ty_i\|^2\right)^{\frac{1}{2}}$$
$$\leq (1 - \|Ty\|^2)^{\frac{1}{2}} .$$

Proceeding to the limit, we have

$$1 - \phi(|y|) \leq (1 - \|Ty\|^2)^{\frac{1}{2}} ,$$

from which it follows that

$$\|Ty\|^2 \leq 2\phi(|y|) - \phi(|y|)^2 \leq 2\phi(|y|) .$$

Now choose any non-zero element x, and put $y = x/\|x\|$. We obtain

$$\|Tx\|^2 \leq 2\phi(|x|)\,\|x\| .$$

The original statement follows at once.

For $q > p$, the same proof gives

$$\|Tx\|^q \leq q\, \pi_{q,p}(T)^q\, \phi(|x|^p)\, \|x\|^{q-p} .$$

Pisier (1986) derives a theorem on factorization of such operators through a suitable "Lorentz function space", and also generalizes the theorem to operators on C*-algebras.

We mention one attractive application of Pisier's theorem:

<u>14.2.</u> For any space E,

$$\pi_1(E) \leqslant 2\, \pi_{2,1}(E)^2\, \lambda(E)^2 .$$

If $\lambda(E)$ and $\pi_{2,1}(E)$ are both finite, then E is finite-dimensional.

Proof. Embed E in a space $\ell_\infty(S)$, and take a projection P with $\|P\|$ close to $\lambda(E)$. By 14.1, there is a function ϕ on $\ell_\infty(S)$ such that $\|\phi\| = 1$ and

$$\|Px\|^2 \leqslant 2\, \pi_{2,1}(P)^2\, \phi(|x|)\, \|x\|$$

for all x. For $x \in E$, we have $Px = x$, and hence

$$\|x\| \leqslant 2\, \pi_{2,1}(P)^2 \phi(|x|) ,$$

from which it follows (by 3.17) that $\pi_1(E) \leqslant 2\pi_{2,1}(P)^2$. Finally, we have $\pi_{2,1}(P) \leqslant \pi_{2,1}(E)\, \|P\|$.

By 11.2, it follows further that $\Delta_2(E) \leqslant \sqrt{2}\, \pi_{2,1}(E)\, \lambda(E)$.

Using the same idea, we can easily show that Pisier's theorem does not apply to operators defined on subspaces of L_∞ :

<u>14.3 Example.</u> Let E_n be an isometric copy of ℓ_1^n in ℓ_∞^N . We know that $\pi_{2,1}(E_n) = \sqrt{2}$. If ϕ is a functional on ℓ_∞^N such that $\|\phi\| = 1$ and $\|x\|^2 \leqslant K^2\phi(|x|)\, \|x\|$ for all $x \in E_n$, then $\pi_1(E_n) \leqslant K^2$, so $K^2 \geqslant \sqrt{n}$.

<u>Exercise.</u> Let T be an operator defined on a subspace E of an L_∞-space X. Show (roughly as in the proof of Pietsch's theorem) that the following statements are equivalent:

(i) There is a positive functional ϕ on X such that $\|\phi\| = 1$ and $\|Tx\|^2 \leqslant C^2\phi(|x|)\, \|x\|$ for all $x \in E$.

(ii) For elements x_i of E,

$$\Sigma\, \|Tx_i\|^2 \leqslant C^2\, \|\, \Sigma\, \|x_i\|\, |x_i|\, \| .$$

Exercise. Adapt the premultiplication lemma 3.6 to $\pi_{2,1}$, and deduce the following statement : if $\pi_{2,1}(T) \leq K\,\|T\|$ for all operators T from ℓ_∞^n to X (for any n), then $\pi_{2,1}(X) \leq K$.

14.4 Proposition. (Maurey, 1974b). Let T be an operator on ℓ_∞^n. Then $\pi_{2,1}(T)$ can be computed using only disjointly supported elements of ℓ_∞^n.

Proof. (We give only a sketch). Let $x_1, \dots, x_k$ be elements of ℓ_∞^n with $\| \Sigma\, |x_i| \,\| = 1$. We show that there exist elements $x_i^{(j)}$, mutually disjoint for each fixed j, such that

$$x_i = \sum_j x_i^{(j)} \quad \text{for each } i,$$

$$\sum_j \|u^{(j)}\| = 1, \quad \text{where} \quad u^{(j)} = \sum_i |x_i^{(j)}| \,.$$

The statement then follows easily.

Introduce a further element (if necessary) to ensure that

$$\sum_i |x_i(r)| = 1 \quad \text{for all } r.$$

For each r, let $\delta(r) = \max_i |x_i(r)| \;\; (\geq \frac{1}{k})$. Let $\delta_1 = \min \delta(r)$. Choose r. For some p, we have $|x_p(r)| \geq \delta_1$. Define $x_p(r)$ to be $\delta_1 \operatorname{sgn} x_p(r)$, and let $x_i(r) = 0$ for $i \neq p$. In this way, we obtain disjoint elements $x_i^{(1)}$ with

$$\sum_i |x_i^{(1)}| = (\delta_1, \dots, \delta_1).$$

Now apply the same process to the elements $x_i - x_i^{(1)}$, and repeat.

14.5 Example. This example (an unpublished one due to S.J. Montgomery-Smith) shows that the extra 2 appearing in Pisier's theorem cannot be simply removed.

Let Y be $\mathbb{R}^3$ with norm

$$\|y\|_Y = \max\, [|y(2)| + |y(3)|,\ |y(3)| + |y(1)|,\ |y(1)| + |y(2)|\,]$$

Let T be the identity mapping from ℓ_∞^3 to Y. We have $\|e_i\| = 1$ and $\|e_i+e_j\|_Y = \|e_1+e_2+e_3\|_Y = 2$. Using 14.4, one deduces easily that $\pi_{2,1}(T)^2 = 5$.

Suppose that $\lambda_1, \lambda_2, \lambda_3 \geq 0$ are such that

$$\|Tx\|^2 \leq \Sigma\, \lambda_i\, |x(i)|\, \|x\|$$

for all x. Then $\lambda_i+\lambda_j \geq 4$ for distinct i,j, hence $\lambda_1+\lambda_2+\lambda_3 \geq 6$.

It is tempting to conjecture that there is a constant C, independent of n, such that $\pi_2(T) \leq C\,\pi_{2,1}(T)$ for all operators on ℓ_∞^n. This would imply that all spaces with the Orlicz property are 2-dominated, and Grothendieck's inequality would follow easily. We finish this short section with an example that disproves this conjecture. It also provides a nice illustration of the use of Pisier's criterion to estimate $\pi_{2,1}(T)$.

<u>14.6 Example.</u> Consider the "cyclic" operator T on ℓ_∞^n defined in 3.29. It was shown there that $\pi_2(T) = \sqrt{n}\,\|a\|_2$.

Suppose now that a is a decreasing, non-negative function, with $a(j) = a_j$. Let ϕ be the functional given by $\phi(x) = \Sigma\, x(i)$, so that $\|\phi\| = n$. Let x be an element of ℓ_∞^n, and write $\phi(|x|) = c\|x\|$. We have $\|Tx\| = |\,\Sigma a_i y_i\,|$, where (y_i) is some permutation of the terms of x. Write $Y_j = y_1 + \dots + y_j$, and let r be the integer such that $r \leq c < r+1$. Then

$$|Y_j| \leq j\|x\| \qquad \text{for } j \leq r,$$

$$|Y_j| \leq \phi(|x|) = c\|x\| \qquad \text{for } j > r.$$

By "Abel summation", we have

$$\begin{aligned}
|\sum_1^n a_i y_i| &= |(a_1-a_2)Y_1 + (a_2-a_3)Y_2 + \dots + a_n Y_n| \\
&\leq (a_1-a_2)\|x\| + 2(a_2-a_3)\|x\| + \dots + r(a_r-a_{r+1})\|x\| + c\,a_{r+1}\|x\| \\
&= [a_1 + \dots + a_r + (c-r)a_{r+1}]\,\|x\| \\
&\leq \|x\| \int_0^c a(t)dt.
\end{aligned}$$

Now let $a(t) = t^{-1/2}$. Then

$$\pi_2(T) = \sqrt{n}\left(1 + \frac{1}{2} + \dots + \frac{1}{n}\right) > (n \log n)^{1/2}.$$

By the above,

$$\|Tx\| \leq 2\|x\|c^{1/2} = 2\,\phi(|x|)^{1/2}\,\|x\|^{1/2}.$$

Since $\|\phi\| = n$, it follows that $\pi_{2,1}(T) \leq 2\sqrt{n}$.

The author (1987) has shown that there is a constant C (independent of k,n) such that for all operators of rank n on ℓ_∞^k (for any k), $\pi_2(T) \leq C(\log n)^{1/2}\, \pi_{2,1}(T)$. In this sense, the ratio of $\pi_2(T)$ to $\pi_{2,1}(T)$ grows "very slowly" with n. The above example shows that the factor $(\log n)^{1/2}$ is needed. Of course, it follows that for any n-dimensional space X, we have $\Delta_2(X) \leq C(\log n)^{1/2}\, \pi_{2,1}(X)$.

<u>Exercise.</u> Show that for any operator T on ℓ_∞^n, we have $\pi_1(T) \leq \sqrt{n}\, \pi_{2,1}(T)$ and $\pi_2(T) \leq n^{1/4}\pi_{2,1}(T)$ (use 3.14 and 11.1).

15. TENSOR PRODUCTS OF OPERATORS

This section does not require any prior knowledge of tensor products. Considerations will be restricted to finite-dimensional spaces.

Recall that for finite-dimensional spaces X,Y, the algebraic tensor products $X \otimes Y$ equates to the space $L(X^*,Y)$. The "injective" tensor product norm ε corresponds to ordinary operator norm on $L(X^*,Y)$, so that

$$\varepsilon(\sum_i x_i \otimes y_i) = \sup \{ |\sum_i f(x_i)y_i| : f \in U_{X^*}\}$$

$$= \sup \{ |\sum_i f(x_i)\, g(y_i)| : f \in U_{X^*},\ g \in U_{Y^*}\} .$$

We write $X \otimes_\varepsilon Y$ for $X \otimes Y$ with this norm; this is the only tensor product norm we will be considering. Given $f \in X^*$ and $g \in Y^*$, the element $f \otimes g$ identifies with a linear functional on $X \otimes_\varepsilon Y$ by : $(f \otimes g)(x \otimes y) = f(x)g(y)$. It is elementary that the norm of this functional is $\|f\|.\|g\|$. Further, the above expression shows that $\{f \otimes g : f \in U_{X^*},\ g \in U_{Y^*}\}$ is a norming set of functionals. (By 1.11, we know in fact that the dual of $X \otimes_\varepsilon Y$ identifies with $X^* \otimes_\gamma Y^*$ in this way). The following easy consequence will be needed:

15.1. Given elements x_i of X, y_i of Y, we have

$$\mu_p \{x_i \otimes y_j : 1 \le i \le k,\ 1 \le j \le \ell\} = \mu_p(x_1, \dots, x_k)\, \mu_p(y_1, \dots, y_\ell)$$

in the space $X \otimes_\varepsilon Y$.

Proof. This follows as once from the preceding remark and the equality:

$$\sum_i \sum_j |(f(x_i)g(y_j)|^p = \sum_i |f(x_i)|^p \sum_j |g(y_j)|^p .$$

Now suppose that we have elements S of $L(X_1,X_2)$ and T of $L(Y_1,Y_2)$. Then an operator $S \otimes T$ from $X_1 \otimes Y_1$ to $X_2 \otimes Y_2$ is defined by:

$$(S \otimes T)(x \otimes y) = Sx \otimes Ty .$$

One can verify as in 1.3 that this definition is consistent. Clearly, $I_X \otimes I_Y$ equals $I_{X \otimes Y}$. Our goal is to show that if the spaces $X_i \otimes Y_i$ are given the norm $\mathcal{E}$, then $\alpha(S \otimes T) = \alpha(S)\alpha(T)$ for our "usual" operator ideal norms α. These results appeared first in Holub (1970) and [FDBS]. We start with some algebraic preliminaries.

15.2 (i) If $S = \sum_i f_i \otimes x_i$ and $T = \sum_j g_j \otimes y_j$, then

$$S \otimes T = \sum_i \sum_j (f_i \otimes g_j) \otimes (x_i \otimes y_j) .$$

(ii) If $S \in L(X)$ and $T \in L(Y)$, then trace $(S \otimes T)$ = (trace S) (trace T).

Proof. (i) Applied to an element of the form $x \otimes y$, we have

$$(Sx) \otimes (Ty) = \Big(\sum_i f_i(x) x_i\Big) \otimes \Big(\sum_j g_j(y) y_j\Big)$$
$$= \sum_i \sum_j f_i(x) g_j(y)(x_i \otimes y_j) .$$

(ii) With S,T as in (i), we have

$$\operatorname{trace}(S \otimes T) = \sum_i \sum_j (f_i \otimes g_j)(x_i \otimes y_j)$$
$$= \sum_i \sum_j f_i(x_i) g_j(y_j)$$
$$= (\operatorname{trace} S)(\operatorname{trace} T) .$$

15.3. Let α be an operator ideal norm, and α^* its dual under finite-dimensional trace duality. Suppose that for all operators between finite-dimensional spaces, we have $\alpha(S_1 \otimes S_2) \leq \alpha(S_1)\,\alpha(S_2)$. Then $\alpha^*(T_1 \otimes T_2) \geq \alpha^*(T_1)\,\alpha^*(T_2)$ for all such operators.

Proof. Choose operators T_1,T_2. There exist operators S_i such that $\alpha(S_i) = 1$ and trace $(T_iS_i) = \alpha^*(T_i)$ for $i = 1,2$. Clearly, $(T_1 \otimes T_2)(S_1 \otimes S_2) = (T_1S_1) \otimes (T_2S_2)$, so

$$\begin{aligned}\operatorname{trace}[(T_1 \otimes T_2)(S_1 \otimes S_2)] &= \operatorname{trace}(T_1S_1)\operatorname{trace}(T_2S_2)\\ &= \alpha^*(T_1)\alpha^*(T_2)\,.\end{aligned}$$

Since $\alpha(S_1 \otimes S_2) \leqslant 1$, the statement follows.

Note: The argument is not reversible, since operators from $Y_1 \otimes Y_2$ to $X_1 \otimes X_2$ are not all of the form $S_1 \otimes S_2$.

We are now ready to start considering particular norms. In each case, the spaces are finite-dimensional, S is in $L(X_1,X_2)$, T is in $L(Y_1,Y_2)$ and $S \otimes T$ is regarded as an operator from $X_1 \otimes_\varepsilon Y_1$ to $X_2 \otimes_\varepsilon Y_2$.

<u>15.4.</u> $\|S \otimes T\| = \|S\|.\|T\|$.

Proof. There exist elements x_0,y_0 with $\|x_0\| = \|y\| = 1$ and $\|Sx_0\| = \|S\|$, $\|Ty_0\| = \|T\|$. Then $\varepsilon(x_0 \otimes y_0) = 1$, while $\varepsilon(Sx_0 \otimes Ty_0) = \|S\|.\|T\|$. Hence $\|S \otimes T\| \geqslant \|S\|.\|T\|$.

Now let $u = \sum_i x_i \otimes y_i$, with $\varepsilon(u) = 1$. Then $(S \otimes T)(u) = \sum_i (Sx_i) \otimes (Ty_i)$. If $f \in X_2^*$ and $g \in Y_2^*$ have $\|f\| = \|g\| = 1$, then

$$\begin{aligned}\Big| \sum_i f(Sx_i)\, g(Ty_i)\Big| &= \Big| \sum_i (S^*f)(x_i)(T^*g)(y_i) \Big|\\ &\leqslant \|S^*f\|.\|T^*g\| \leqslant \|S\|.\|T\| .\end{aligned}$$

This shows that $\varepsilon[(S \otimes T)(u)] \leqslant \|S\|.\|T\|$, and hence $\|S \otimes T\| \leqslant \|S\|.\|T\|$.

<u>15.5.</u> $\nu_1(S \otimes T) = \nu_1(S)\nu_1(T)$.

Proof. It follows at once from the expression for $S \otimes T$ in 15.2 that $\nu_1(S \otimes T) \leqslant \nu_1(S)\nu_1(T)$.

Since $\nu_1^* = \|\ \|$, the opposite inequality follows from 15.3 and 15.4.

In the same way, we now obtain similar results for π_1 and ν_∞. These cases are especially interesting, since they already say something non-trivial for identity operators.

<u>15.6 Proposition.</u> We have $\pi_1(S \otimes T) = \pi_1(S)\pi_1(T)$ and $\nu_\infty(S \otimes T) = \nu_\infty(S)\,\nu_\infty(T)$.

In particular, $\pi_1(X \otimes_\varepsilon Y) = \pi_1(X)\pi_1(Y)$ and $\lambda(X \otimes_\varepsilon Y) = \lambda(X)\lambda(Y)$.

Proof. We shall show that $\nu_\infty(S \otimes T) \leq \nu_\infty(S)\nu_\infty(T)$ and $\pi_1(S \otimes T) \leq \pi_1(S)\pi_1(T)$. Since $\nu_\infty^* = \pi_1$, the statements then follow by 15.3.

To prove the inequality for ν_∞, choose representations $S = \Sigma f_i \otimes x_i$, $T = \Sigma g_j \otimes y_j$ with $\|f_i\| = \|g_j\| = 1$ and $\mu_1(x_1, \ldots, x_k)$, $\mu_1(y_1, \ldots, y_\ell)$ close to $\nu_\infty(S)$, $\nu_\infty(T)$ respectively. The required inequality follows from the expression in 15.2, together with 15.1.

We now consider π_1. Take $\delta > 0$. By 5.3, there are functionals f_i, g_j such that

$$\|Sx\| \leq \sum_i |f_i(x)| \quad \text{for } x \in X_1, \qquad \sum_i \|f_i\| \leq (1+\delta)\pi_1(S) ,$$

$$\|Ty\| \leq \sum_j |g_j(y)| \quad \text{for } y \in Y_1, \qquad \sum_j \|g_j\| \leq (1+\delta)\pi_1(T) .$$

We show that $\varepsilon[(S \otimes T)(u)] \leq \sum_i \sum_j |(f_i \otimes g_j)(u)|$ for all u in $X_1 \otimes Y_1$; the required inequality then follows. Write $u = \sum_r x_r \otimes y_r$ and $v = (S \otimes T)(u) = \sum_r (Sx_r) \otimes (Ty_r)$. There is an element h of X_2^* with $\|h\| = 1$ and

$$\varepsilon(v) = \|\sum_r h(Sx_r)(Ty_r)\| = \|T(\sum_r h(Sx_r)y_r)\|$$

By the choice of the g_j, this is not greater than

$$\sum_j |\sum_r h(Sx_r)\, g_j(y_r)| = \sum_j |h(z_j)| \leq \sum_j \|z_j\| ,$$

where $z_j = \sum_r g_j(y_r)(Sx_r)$. By the choice of the f_i ,

$$\|z_j\| \leq \sum_i |\sum_r f_i(x_r)\, g_j(y_r)| = \sum_i |(f_i \otimes g_j)(u)| .$$

The stated inequality now follows.

It is also easy to prove directly that $\pi_1(S \otimes T) \geq \pi_1(S)\, \pi_1(T)$, without using ν_∞.

So we are now in a position to state exact values of π_1 and λ for spaces of the form $L(\ell_p^m, \ell_q^n)$. For example, the projection constant of $L(\ell_2^n)$ is $\lambda(\ell_2^n)^2$ (see 8.8, 8.10 for the value of $\lambda(\ell_2^n)$).

<u>15.7.</u> $\pi_2(S \otimes T) = \pi_2(S)\, \pi_2(T)$.

Proof. Recall that ν_2 coincides with π_2 and $\nu_2^* = \pi_2$. From 15.1 and 15.2 one has $\nu_2(S \otimes T) \leq \nu_2(S)\nu_2(T)$. Equality follows, by 15.3.

<u>15.8 Example.</u> It is elementary to verify (using 15.1 and the definition) that $\pi_{2,1}(S \otimes T) \geq \pi_{2,1}(S)\ \pi_{2,1}(T)$. However, equality does not hold in this case. For instance, $Z = \ell_2^n \otimes_\varepsilon \ell_2^n$ contains an isometric copy of ℓ_∞^n (the diagonal operators), so $\pi_{2,1}(Z) \geq \sqrt{n}$. Of course, $\pi_{2,1}(\ell_2^n) = 1$.

We finish this section by remarking briefly on what happens when $X \otimes_\varepsilon Y$ is replaced by $X \otimes_\gamma Y$ (that is, $L(X^*,Y)$ with norm ν_1). It is a very straightforward exercise to show that we still have $\|S \otimes T\| = \|S\|.\|T\|$ and $\nu_1(S \otimes T) = \nu_1(S)\nu_1(T)$. However, equality no longer applies in the other cases. For π_1 and ν_∞, this can be seen from the values given for ℓ_1^n in section 7. It follows from 1.14 that $\ell_1^m \otimes_\gamma \ell_1^n$ is isometric to ℓ_1^{mn} , and in general $\lambda(\ell_1^{mn})$ is different from $\lambda(\ell_1^m)\lambda(\ell_1^n)$.

16. TRACE DUALITY REVISITED : INTEGRAL NORMS

Second duals under trace duality

Under trace duality, the summing norms are the duals of the nuclear norms, but we have not yet given any description of the duals of the summing norms. A related problem is as follows. We know (6.2) that $\nu_1(I_{1,\infty}^{(n)}) = 1$ for all n. Is there a corresponding statement for the infinite-rank operator $I_{1,\infty}$? In both cases, what is needed is an extension of the concept of the nuclear norms that is not confined to finite-rank (or "nuclear") operators. We now describe a very simple construction that achieves this.

Let α be any operator ideal norm, defined at least for finite-rank operators, such that $\alpha(T) \geq \|T\|$ for all T (in particular, any ν_p). Let X,Y be any normed linear spaces. Define

$$\tilde{\alpha}(T) = \sup\{\alpha(T|_{X_0}) : X_0 \text{ a finite-dimensional subspace of } X\}$$

for all T in L(X,Y) for which this is finite. Denote the set of such T by $L_{\tilde{\alpha}}(X,Y)$. One checks easily that $\tilde{\alpha}$ is a norm (in fact, an operator ideal norm) on $L_{\tilde{\alpha}}(X,Y)$, with $\tilde{\alpha}(T) \geq \|T\|$. If T is of finite rank, then $\tilde{\alpha}(T) \leq \alpha(T)$, and if X is finite-dimensional, then $\tilde{\alpha}(T) = \alpha(T)$.

Note that $T|_{X_0}$ can be written as TJ, where J is the inclusion operator $X_0 \to X$. If X_1 is any finite-dimensional space, and A is an operator from X_1 to X with $\|A\| \leq 1$, then $TA = (T|_{X_0})A$, where $X_0 = A(X_1)$. Hence $\alpha(TA) \leq \tilde{\alpha}(T)$. This shows that $\tilde{\alpha}(T)$ can also be described as the supremum of $\alpha(TA)$ over all such X_1 and A.

Recall our definition of the dual norm α^* :

$$\alpha^*(S) = \sup\{|\mathrm{trace}\,(ST)| : T \in FL(X,Y),\ \alpha(T) \leq 1\}$$

for operators S from Y to X. It is trivial to verify that $\alpha^*(AS) \leq \|A\|\alpha^*(S)$

for an operator A on X (note that trace(AS.T) = trace(S.TA)). We say that α^* is injective if (like the summing norms) it does not depend on the range space : in other words, if $S(Y) \subseteq X_0 \subseteq X_1$ and S_0 is the same operator regarded as an element of $L(Y,X_0)$, then $\alpha^*(S_0) = \alpha^*(S)$.

16.1 Proposition. If α is an operator ideal norm such that α^* is injective, then α^{**} coincides with $\tilde{\alpha}$.

Proof. Suppose that T is an element of $L(X,Y)$ with $\tilde{\alpha}(T)$ finite. Let S be an element of $FL(Y,X)$ with $\alpha^*(S) = 1$. Write $X_0 = S(Y)$. Let S_0 be S, regarded as an element of $L(Y,X_0)$, and let J be the inclusion operator $X_0 \to X$, so that $S = JS_0$. By hypothesis, $\alpha^*(S_0) = \alpha^*(S) = 1$, so

$$|\text{trace } (TS)| = |\text{trace } (TJ.S_0)| \leq \alpha(TJ) \leq \tilde{\alpha}(T) .$$

Hence $\alpha^{**}(T) \leq \tilde{\alpha}(T)$.

Now suppose that $\alpha^{**}(T)$ is finite. Let X_1 be finite-dimensional, and let A be an element of $L(X_1,X)$ with $\|A\| = 1$. By 1.8, the Banach space dual of $[L(X_1,Y),\alpha]$ identifies with $[L(Y,X_1),\alpha^*]$, so there is an element S of $L(Y,X_1)$ with $\alpha^*(S) = 1$ and trace $(TA.S) = \alpha(TA)$. Then $\alpha^*(AS) \leq 1$, so trace $(T.AS) \leq \alpha^{**}(T)$. Hence $\tilde{\alpha}(T) \leq \alpha^{**}(T)$.

The integral norms

We now define the p-integral norm i_p to be $\tilde{\nu}_p$. We say that an operator T is "p-integral" if $i_p(T)$ is finite. So $i_p = \pi_{p'}{}^*$, where $\frac{1}{p} + \frac{1}{p'} = 1$; also, $i_1 = \| \ \|^*$ and $i_\infty = \pi_1{}^*$.

Clearly, $i_p(T) \leq \nu_p(T)$ for finite-rank operators, with equality when the domain is finite-dimensional. One might expect equality to hold generally, but this question is not as simple as it sounds. We will show in Section 17 that equality holds when the range space Y is finite-dimensional. It is elementary that $|\text{trace } T| \leq i_1(T)$ for all T in $FL(X)$.

Since $\pi_p(T) \leq \nu_p(T)$ for finite-rank operators, we have $\pi_p(T) \leq i_p(T)$ whenever T is p-integral.

With the integral norms at our disposal, we can very easily formulate new versions of some of our earlier results involving the nuclear norms, free of the restriction to finite-rank operators. The case p = 2 is particularly simple:

<u>16.2</u> The classes of 2-summing and 2-integral operators coincide, and i_2 coincides with π_2.

Proof. Let T be an operator on X. For every finite-dimensional subspace X_0 of X, we have (by 5.11) $\nu_2(T|_{X_0}) = \pi_2(T|_{X_0})$. The statement follows at once.

Similarly, 4.2 translates into the following:

<u>16.3</u> Suppose that S is a p-integral operator from X to Y, and T is a p'-summing operator from Y to Z, where $1/p + 1/p' = 1$. Then $i_1(TS) \leq \pi_{p'}(T)\, i_p(S)$. (If $p = \infty$, then $p' = 1$).

Proof. By 4.2, for any finite-dimensional subspace X_0 of X, we have $\nu_1(TS|_{X_0}) \leq \pi_{p'}(T)\nu_p(S|_{X_0}) \leq \pi_{p'}(T)i_p(S)$.

<u>Exercise.</u> Use 4.5 to show that $i_1(TS) \leq i_p(T)\pi_{p'}(S)$.

Corresponding to 4.6, 5.4, 5.5, we obtain:

<u>16.4.</u> Let X be an $\mathcal{L}_\infty$-space. For all operators from or into X, we have $i_p(T) = \pi_p(T)$ (and $i_\infty(T) = \|T\|$).

Proof. First, consider an operator defined on X. Let X_0 be a finite-dimensional subspace of X_1 and choose $\varepsilon > 0$. There is a subspace X_1 containing X_0 such that $d(X_1, \ell_\infty^N) \leq 1 + \varepsilon$ for some N. By 5.4

$$\nu_p(T|_{X_1}) \leq (1 + \varepsilon)\pi_p(T|_{X_1}) \leq (1 + \varepsilon)\pi_p(T)$$

(and similarly for ν_∞, from 4.6). The statement follows.

Now consider an operator from Y into X. Let Y_0 be a finite-dimensional subspace of Y, and choose $\varepsilon > 0$. Then $T(Y_0)$ is contained in a subspace X_0 of X satisfying $d(X_0, \ell_\infty^N) \leq 1 + \varepsilon$ for some N. Write $T|_{Y_0} = T_0$, and let $\overline{T}_0$ be the same mapping regarded as an element of $L(Y_0, X_0)$. By 5.5,

$$\nu_p(T_0) \leq \nu_p(\overline{T}_0) \leq (1 + \varepsilon)\pi_p(\overline{T}_0) \leq (1 + \varepsilon)\pi_p(T),$$

(and similarly for ν_∞, from 4.7).

In particular, since $\pi_1(I_{1,\infty}) = 1$ (see 6.2), we have $i_1(I_{1,\infty}) = 1$, which answers the question posed at the beginning of this section. Furthermore, we can now substitute i_p for π_p in results applying to operators on $\mathcal{L}_\infty$-spaces (for example, 3.17, 3.18, 3.20).

Finally, note that if X is a 2-dominated space, then $\pi_2(T) \leqslant \Delta_2(X)\, i_\infty(T)$ for operators into X.

An alternative trace duality theory

What we have outlined above is the simplest way of defining the integral norms, and it is perfectly suited for the immediate conversion of the earlier results involving ν_p. However, in the context of the general theory of "operator ideals", the notions $\tilde{\alpha}$ and α^* are not fully satisfactory, since their definition is unsymmetrical and 16.1 requires the extra condition that α^* is injective. To overcome this, one constructs a rather more elaborate theory, as follows. Define $\alpha^{max}(T)$ to be the supremum of $\alpha(BTA)$ for all operators A,B from and into finite-dimensional spaces (respectively) with $\|A\| = \|B\| = 1$. The basic idea is to ensure that one is always working with operators between finite-dimensional spaces. In the same spirit, one defines the "adjoint" norm α^{adj} as follows. For S in L(Y,X), $\alpha^{adj}(S)$ is the supremum of $|\text{trace}\,(BTA.S)|$ taken over the following situations : X_1,Y_1 are finite-dimensional, $A \in L(X,X_1)$, $T \in L(X_1,Y_1)$, $B \in L(Y_1,Y)$ and $\|A\| = \|B\| = \alpha(S) = 1$.

It is elementary to verify that, with no extra conditions, α^{max} coincides with $(\alpha^{adj})^{adj}$; the reader may care to attempt this as an exercise. A very thorough account of these and related ideas is given in [OI] (Pietsch uses the notation α^* for what we have just called α^{adj}).

When this approach is followed, it is natural to define i_p to be ν_p^{max}. It is not too hard to show that this agrees with our definition in the case p = 1. We continue to use the notation i_p as originally defined, that is $\tilde{\nu}_p$. Of course, $\nu_p^{max}(T) \leqslant i_p(T)$.

<u>16.5.</u> ν_1^{max} coincides with i_1 .

Proof. Let T be in L(X,Y), and take $\varepsilon > 0$. By 16.1, there is an element S of FL(Y,X) with $\|S\| = 1$ and trace $(ST) \geqslant (1-\varepsilon)i_1(T)$. Let ker S = K, and write $Y_1 = Y/K$. Then Y_1 is finite-dimensional. Let Q be the quotient mapping of Y onto Y_1. In the standard way, S factorizes as S_1Q, with $\|S_1\| = 1$. Regard S_1 as an operator onto S(Y), and let J be the inclusion operator $S(Y) \to X$. Then $S = JS_1Q$, so

$$\text{trace}\ (TS) = \text{trace}\ (T.JS_1Q) = \text{trace}\ (QTJ.S_1) \leqslant \nu_1(QTJ).$$

This shows that $\nu_1^{max}(T) \geqslant (1 - \varepsilon) i_1(T)$.

The equivalence of the definitions for other p is not trivial; it can be proved from the factorization theorem quoted below. However, a mild extra condition makes it quite easy. The space Y is said to have the metric approximation property if, given $\varepsilon > 0$ and a finite-dimensional subspace Y_0, there is an operator V in FL(Y) such that $Vy = y$ for all $y \in Y_0$ and $\|V\| \leqslant 1 + \varepsilon$. All the "usual" spaces have this property, and in fact it took mathematicians many years to establish the existence of a space without the property. Suppose now that Y has this property, and let T be an operator from a finite-dimensional space into Y. There is an operator V as above, with $Y_0 = T(X)$. Then $T = VT$. Write $V = JV_0$, where V_0 is regarded as an operator into V(Y), and J is the inclusion mapping $V(Y) \to Y$. Then $T = JV_0T$, so $\nu_p(T) \leqslant \nu_p(V_0T)$. It is now easy to deduce that $\nu_p^{max}(T) = i_p(T)$ for all operators into Y.

Exercise. Show that $\pi_p^{max} = \pi_p$ and $\nu_p^{adj} = \pi_p$. (Embed the range of TA in ℓ_∞^N).

Further developments

Let μ be a probability measure on a set S, and let K_p be the natural "identity" mapping from $L_\infty(\mu)$ to $L_p(\mu)$. Then $\pi_p(K_p) = 1$ (this is essentially 3.18), so by 16.4, $i_p(K_p) = 1$. The factorization theorem of Persson & Pietsch (1969) says, roughly speaking, that all p-integral oplerators are of the form BK_pA. The exact statement is :

Theorem. Let T be a p-integral operator from X to Y_1 and let J_Y be the embedding of Y into Y^{**}. Then there is a probability space (S,μ) such that J_YT factorizes as BK_pA, where K_p is as above, A is in $L(X,L_\infty(\mu))$ and B is in $L(L_p(\mu),Y^{**})$. Further, $i_p(T)$ is the infimum of $\|B\|.\|A\|$ taken over all such factorizations.

The corresponding "discrete" factorization result for ν_p (through ℓ_∞^k and ℓ_p^k) was described in 4.11 and the ensuing exercise. The above theorem can be deduced from this, using the notion of "ultraproducts" (see [OI], chapter 19). Note that the case $p = 2$ is essentially Pietsch's theorem.

Some writers prefer to regard the description given by this theorem as the definition of p-integral operators (and norms).

It is a relatively short step from the factorization theorem to the representation of T in the form

$$Tx = \int f(x)\, dM(f)$$

where M is a vector-valued measure on the unit ball of X^*. This in fact, was taken as the initial definition by Persson and Pietsch (following Grothendieck), and it explains (at last!) the term "p-integral." However, vector measures are not particularly helpful for most applications, and this approach is no longer followed by many writers.

17. APPLICATIONS OF LOCAL REFLEXIVITY

The local reflexivity theorem

The following theorem is known as the "principle of local reflexivity." It has many applications; here we describe those relating to the concepts studied in this book.

Theorem. Let X be a normed linear space. Regard X as a subspace of X^{**}. Let E be a finite-dimensional subspace of X^{**}. Let $\varepsilon > 0$, and let elements $g_1, \dots, g_k$ of X^* be given. Then there is an operator $R : E \to X$ such that:

(i) $\|R\| \leq 1+\varepsilon$, $\|R^{-1}\| \leq 1+\varepsilon$,

(ii) $Rz = z$ for z in $E \cap X$,

(iii) $g_i(R\phi) = \phi(g_i)$ for all i and all $\phi \in E$.

Statements (i) and (ii) were obtained by Lindenstrauss and Rosenthal in 1969. Statement (iii) was added by Johnson et al. (1971). For a proof, we refer to this paper and [CBS I].

Applications to dual spaces and operators

Statement (i) of the theorem says that X^{**} is finitely represented in X. As a first application, this gives at once:

17.1. The type 2, cotype 2 and 2-dominated constants have the same value for X^{**} and for X.

Next, we show that the nuclear norms are unchanged if the range is regarded as Y^{**} instead of Y.

<u>17.2.</u> Let T be an element of FL(X,Y), and let J_Y be the embedding of Y into Y^{**}. Then $\nu_p(J_Y T) = \nu_p(T)$ for all p.

Proof. Take $\varepsilon > 0$. Express $J_Y T$ as $\Sigma\, f_i \otimes \phi_i$, where $f_i \in X^*$, $\phi_i \in Y^{**}$, $\Sigma\, \|f_i\|^p = 1$ and $\mu_{p'}(\phi_1, \ldots, \phi_k) \leqslant (1+\varepsilon)\nu_p(J_Y T)$. Let E be the subspace of Y^{**} spanned by the ϕ_i, and let R be the operator in L(E,X) given by local reflexivity. Since $J_Y T$ maps into $E \cap X$, we have $T = \Sigma\, f_i \otimes (R\phi_i)$. The statement follows, since

$$\mu_{p'}(R\phi_1, \ldots, R\phi_k) \leqslant (1+\varepsilon)\mu_{p'}(\phi_1, \ldots, \phi_k) .$$

<u>17.3 Proposition.</u> Let T be any finite-rank operator. Then $\nu_p(T^{**}) = \nu_p(T)$ for all p, and $\nu_1(T^*) = \nu_1(T)$.

Proof. We know from the expression in 1.2 that $\nu_p(T^{**}) \leqslant \nu_p(T)$. With the notation of 17.2, $J_Y T$ is the restriction of T^{**} to X. Hence, by 17.2,

$$\nu_p(T) = \nu_p(J_Y T) \leqslant \nu_p(T^{**}) .$$

The statement for p = 1 follows, since $\nu_1(T^*) \leqslant \nu_1(T)$.

Using statement (iii) in the local reflexivity theorem, we can prove a similar result for the summing norms.

<u>17.4 Proposition.</u> For any p-summing operator T, we have $\pi_p(T^{**}) = \pi_p(T)$.

Proof. Since T^{**} is an "extension" of T (when X is regarded as a subspace of X^{**}), we have $\pi_p(T^{**}) \geqslant \pi_p(T)$.

Choose elements $\phi_1, \ldots, \phi_k$ of X^{**} such that $\mu_p(\phi_1, \ldots, \phi_k) = 1$ and

$$\left(\Sigma\, \|T^{**}\phi_i\|^p \right)^{1/p} \geqslant (1-\varepsilon)\, \pi_p(T^{**}) .$$

For each i, choose g_i in U_{Y^*} such that

$$(1-\varepsilon)\, \|T^{**}\phi_i\| \leqslant (T^{**}\phi_i)(g_i) = \phi_i(T^* g_i) .$$

By local reflexivity, there is an operator R from $\mathrm{lin}(\phi_1, \ldots, \phi_k)$ to X such that $\|R\| \leqslant 1+\varepsilon$ and $(T^* g_i)(R\phi_i) = \phi_i(T^* g_i)$ for each i. Write $x_i = R\phi_i$. Then $\mu_p(x_1, \ldots, x_k) \leqslant 1+\varepsilon$ and

$$\|Tx_i\| \geq g_i(Tx_i) = (T^*g_i)(x_i) = \phi_i(T^*g_i) \geq (1-\varepsilon)\|T^{**}\phi_i\| .$$

It follows that $\pi_p(T) \geq \pi_p(T^{**})$.

The same reasoning applies to the mixed summing norms (in particular $\pi_{2,1}$). Hence if X has the Orlicz property, so does X^{**}.

Applications to trace duality

The starting point for this is a slight generalization of our original result (1.8) identifying functionals with operators by trace duality. For any A in $FL(X,X^{**})$, one can still define trace A : if A is $\Sigma f_i \otimes \phi_i$, with $f_i \in X^*$ and $\phi_i \in X^{**}$, then its trace is $\Sigma \phi_i(f_i)$.

If we now have an element T of $L(Y,X^{**})$, then a linear functional ϕ_T on $FL(X,Y)$ is defined by $\phi_T(S)$ = trace (TS) (not trace (ST) !). Conversely, if α is an operator ideal norm as $FL(X,Y)$ and ϕ is a continuous functional on $[FL(X,Y),\alpha]$, then the proof of 1.8 shows that ϕ is ϕ_T for some such T : in fact, Ty is the functional on X^{**} defined by

$$(Ty)(f) = \phi(f \otimes y).$$

Finally, the proofs of 1.11 and 4.3 still apply (with trivial changes) to show that if α is ν_p, then the norm of the functional ϕ_T is $\pi_{p'}(T)$. So we can state :

17.5. The Banach space dual of $[FL(X,Y),\nu_p]$ equates with $[L(Y,X^{**}), \pi_{p'}]$, where $1/p + 1/p' = 1$ (and $\pi_{p'} = \|\ \|$ when $p = 1$).

We now apply the local reflexivity theorem to the X^{**} that has emerged here. We use the notation for integral norms introduced in section 16 : for S in $L(X,Y)$,

$$i_p(S) = \sup \{\nu_p(S|_E) : E \text{ finite-dimensional}\}$$

$$= \sup \{ |\text{trace}(TS)| : T \in FL(Y,X), \pi_{p'}(T) \leq 1\} .$$

17.6 Proposition. Suppose that Y is finite-dimensional and S is an element of $L(X,Y)$. Then $\nu_p(S) = i_p(S)$ for each p.

Proof. Choose some representation $\Sigma f_i \otimes y_i$ for S. By 17.5, there is an element T of $L(Y,X^{**})$ such that $\pi_{p'}(T) = 1$ and

$$\nu_p(S) = \text{trace }(TS) = \Sigma\,(Ty_i)(f_i)\ .$$

Take $\varepsilon > 0$. By local reflexivity, there is an operator $R : T(Y) \to X$ such that $\|R\| \leqslant 1+\varepsilon$ and

$$f_i(RTy_i) = (Ty_i)(f_i)$$

for each i. Let $T_1 = RT$. Then T_1 is in L(Y,X) and $\pi_{p'}(T_1) \leqslant \|R\|\ \pi_{p'}(T) \leqslant 1+\varepsilon$. Further,

$$\text{trace }(T_1S) = \Sigma\, f_i(T_1y_i) = \Sigma\,(Ty_i)(f_i) = \text{trace }(TS)\ .$$

This completes the proof.

This enables us to extend 5.11 to the fully general case:

<u>17.7 Proposition.</u> For any finite-rank operator S, $\nu_2(S) = \pi_2(S)$.

Proof. Let S be in FL(X,Y). Let $Y_1 = S(X)$, and let S_1 be the same mapping regarded as an element of $FL(X,Y_1)$. Then $\nu_2(S) \leqslant \nu_2(S_1)$. By 17.6 and 16.2, $\nu_2(S_1) = i_2(S_1) = \pi_2(S_1)$. But $\pi_2(S_1) = \pi_2(S)$, so the statement follows.

Of course, it follows that ν_2 is independent of the stated range space (i.e. it is "injective").

We stress that 17.6 does not extend in the same easy way to all finite-rank operators.

<u>Exercise.</u> If T is p-integral and B has finite rank, show that $\nu_p(BT) \leqslant \|B\|\ i_p(T)$.

<u>Exercise.</u> If S is in $P_{p'}(X,Y)$ and T is in FL(Y,Z), prove that $\nu_1(TS) \leqslant \nu_p(T)\ \pi_{p'}(S)$, where $1/p + 1/p' = 1$ (compare 4.5).

18. CONE-SUMMING NORMS

Elementary theory

Ideas connected with positivity have permeated a good deal of the work in this book. For operators defined on a normed lattice, it is natural to consider a "summing" norm that is defined in a way that pays attention to the order structure. The simplest way to do this is to restrict to positive elements in the definition of π_1. The resulting "cone-summing" norm gives rise to a theory that parallels closely (and in places more simply) the most successful parts of the theory of 1-summing and 2-summing norms. It also provides a proper setting for our sporadic earlier remarks on positive operators. The concept and the basic results are due to Schlotterbeck (1971).

To set the scene, we need a few very elementary concepts and results relating to normed lattices. The definition was given in Section 0. The set $\{x : x \geqslant 0\}$ in a linear lattice X is called the positive cone, and will be denoted by X_+. The supremum of the two elements x,y is denoted by $x \vee y$, the infimum by $x \wedge y$. It is clear that

$$(x+z) \vee (y+z) = (x \vee y) + z \tag{1}$$

We shall use this repeatedly. We write $x^+ = x \vee 0$, $x^- = (-x) \vee 0$, $|x| = x \vee (-x)$. Clearly, $|\lambda x| = |\lambda|.|x|$ and $|x+y| \leqslant |x| + |y|$. Further:

18.1. For any x, we have $x = x^+ - x^-$ and $|x| = x^+ + x^-$.

Proof. We have by (1)

$$x + 0 \vee (-x) = x \vee 0,$$

that is, $x + x^- = x^+$. Further,

$$|x| = x \vee (-x) = (2x \vee 0) - x$$

$$= 2x^+ - (x^+ - x^-) = x^+ - x^-.$$

This decomposition is the key to most of what follows. Notice that it implies that $\|x^+\| \leqslant \|x\|$.

<u>18.2.</u> Let T be an operator on a normed lattice. Write

$$q(T) = \sup \{ \|Tx\| : x \in U_X \cap X_+ \} .$$

Then $q(T) \geqslant \frac{1}{2} \|T\|$. If T is a positive operator (into another normed lattice), then $q(T) = \|T\|$.

Proof. If $\|x\| \leqslant 1$, then $\|x^+\| \leqslant 1$ and $\|x^-\| \leqslant 1$. Since $Tx = Tx^+ - Tx^-$, it follows that $\|T\| \leqslant 2q(T)$.

If T is positive, then for any x, we have $|Tx| \leqslant T(|x|)$, hence $\|Tx\| \leqslant \|T(|x|)\|$. It follows easily that $q(T) = \|T\|$.

The factor $\frac{1}{2}$ is needed. For example, for the functional on ℓ_∞^2 defined by $f(x) = x(1) - x(2)$, we have $\|f\| = 2$, while $q(f) = 1$.

The dual question is resolved by the following lattice version of the Hahn-Banach theorem.

<u>18.3.</u> (i) Let E be a linear sublattice of a normed lattice X, and let f_0 be a continuous, positive functional defined on E. Then there is a continuous, positive functional f on X that extends f_0 and has the same norm.

(ii) If a is an element of a normed lattice, then

$$\sup \{ |f(a)| : f \in X_+^*, \|f\| \leqslant 1\} = \max (\|a^+\|, \|a^-\|).$$

Proof. A sublinear function p is defined on X by putting $p(x) = \|x^+\|$. Then $p(x) = \|x\|$ for $x \geqslant 0$, while $p(x) = 0$ for $x \leqslant 0$. So if f is a linear functional, then the statement that $f(x) \leqslant p(x)$ for all $x \in X$ is equivalent to f being positive and $\|f\| \leqslant 1$. Both statements follow at once from the Hahn-Banach theorem (in the case of (ii), applied to a and -a).

Clearly, $\max (\|a^+\|, \|a^-\|) \geqslant \frac{1}{2} \|a\|$. The element (1,-1) of ℓ_1^2 shows that the factor $\frac{1}{2}$ is needed. Of course, for a *positive* element a, (ii) says that there is a positive functional f such that $\|f\| = 1$ and $f(a) = \|a\|$.

The dual of a normed lattice is itself a normed lattice. For $f \in X^*$, the elements f^+ and $|f|$ are defined (for $x \in X_+$) by:

$$f^+(x) = \sup \{f(y) : 0 \leqslant y \leqslant x\} ,$$

$$|f| (x) = \sup \{f(z) : |z| \leqslant x\}.$$

Two important special types of lattices are as follows. A normed lattice is said to be an L-space if $\|x+y\| = \|x\| + \|y\|$ for all positive elements x,y, and an M-space if $\|x \vee y\| = \max(\|x\|, \|y\|)$ for such elements. (Completeness is often included in the definition, but we will specify this condition when it is needed). Examples of L-spaces are ℓ_1^n, ℓ_1, $L_1(\mu)$, and examples of M-spaces are ℓ_∞^n, ℓ_∞, C(K). (Actually, Kakutani's representation theorem identifies every complete L-space with a space $L_1(\mu)$, and every M-space with a sublattice of some C(K)). It is quite easy to prove that the dual of an L-space is an M-space and conversely.

If X is an L-space, then $\|x\| = \|x^+ + x^-\| = \|x^+\| + \|x^-\|$ for all x, and hence, with the notation of 18.3, $q(T) = \|T\|$ for all operators T on X.

We saw in 2.10 that for positive elements of a normed lattice, $\mu_1(x_1, \dots, x_k) = \|\Sigma x_i\|$. More generally, we have:

<u>18.4.</u> For any elements x_i of a normed lattice, $\mu_1(x_1, \dots, x_k) \leq \| \Sigma |x_i| \|$. Equality holds in an M-space.

Proof. The stated inequality follows from the fact that if $|\alpha_i| \leq 1$ for each i, then $| \Sigma\alpha_i x_i| \leq \Sigma |x_i|$.

Of course, we know from 2.6 that equality holds in the "concrete" M-spaces ℓ_∞, C(K). The reader may be content with this, but we now show how to prove the same for a general M-space (without relying on Kakutani's theorem). The statement follows from the identity

$$\sup \{ | \Sigma \varepsilon_i x_i | : \varepsilon \in D_k\} = \sum_1^k |x_i| ,$$

which we now prove by induction. First, we prove it for k = 2. Choose elements x,y. From (1), we have $|x| + |y| = (|x| + y) \vee (|x| - y)$. By (1) again, $|x| + y = (x+y) \vee (-x+y)$ and $|x| - y = (x-y) \vee (-x-y)$. Hence

$$|x| + |y| = (x+y) \vee (-x-y) \vee (x-y) \vee (-x+y) = |x+y| \vee |x-y| ,$$

as required. Now assume the identity for a certain k, and taken $x_1, \dots, x_{k+1}$. For ε in D_k, write $a_\varepsilon = \sum_1^k \varepsilon_i x_i$. By the case k = 2,

$$|a_\varepsilon + x_{k+1}| \vee |a_\varepsilon - x_{k+1}| = |a_\varepsilon| + |x_{k+1}| .$$

The identity for k+1 follows easily.

Now let T be an operator from X to Y, where X is a normed lattice and Y any normed linear space. The <u>cone-summing norm</u> $\pi^+(T)$ is defined as follows:

$$\pi^+(T) = \sup \{ \Sigma \|Tx_i\| : x_i \in X_+ \text{ and } \|\Sigma x_i\| \leq 1 \},$$

in which all finite sequences (x_i) are considered. The operator T is said to be "cone-summing" if $\pi^+(T)$ is finite.

This definition is what is obtained if attention is restricted to positive elements when defining π_1. It makes sense when X is any partially ordered normed linear space, but we will restrict attention to normed lattices. The notation should really be π_1^+, but we will not attempt to discuss π_p^+ for other p. Some immediate properties are gathered together in the next result.

<u>18.5</u> (i) π^+ is a norm on the space of cone-summing operators from X to Y, and $\|T\| \leq \pi^+(T) \leq \pi_1(T)$.

(ii) If $\| \Sigma|x_i| \| \leq 1$, then $\Sigma \|Tx_i\| \leq \pi^+(T)$.

(iii) For any B in L(Y,Z), we have $\pi^+(BT) \leq \|B\| \pi^+(T)$.

(iv) If V is a normed lattice, and A is a positive operator from V to X, then $\pi^+(TA) \leq \pi^+(T) \|A\|$.

(v) If X and Y are normed lattices and $-T \leq S \leq T$, then $\pi^+(S) \leq \pi^+(T)$.

(vi) If there is a (positive) functional f on X such that $\|Tx\| \leq f(x)$ for all x in X_+, then $\pi^+(T) \leq \|f\|$.

(vii) If X is an L-space, then $\pi^+(T) = \|T\|$. The same is true if Y is an L-space and T is positive.

(viii) If X is an M-space, then $\pi^+(T) = \pi_1(T)$.

Proof. (i),(ii) It is elementary that π^+ is a seminorm. Now $\Sigma |x_i| = \Sigma x_i^+ + \Sigma x_i^-$, so if $\| \Sigma |x_i| \| \leq 1$, then $\Sigma \|Tx_i^+\| + \Sigma \|Tx_i^-\| \leq \pi^+(T)$. Statement (ii) follows, since $\|Tx_i\| \leq \|Tx_i^+\| + \|Tx_i^-\|$. This shows also that $\pi^+(T) \geq \|T\|$, and it follows from 2.10 that $\pi^+(T) \leq \pi_1(T)$.

(iii), (iv) Obvious.

(v) For x_i in X_+, we have $|Sx_i| \leq Tx_i$, hence $\|Sx_i\| \leq \|Tx_i\|$.

(vi) If $x_i \in X_+$ and $\| \Sigma x_i\| \leq 1$, then

$$\Sigma \|Tx_i\| \leq \Sigma f(x_i) \leq \|f\|.$$

(vii) is easy, and (viii) follows from 18.4.

Note that $\pi^+(I_X) = 1$ if and only if X is an L-space. Also, it is now clear that 3.20 arises from the fact that, for a positive operator from an M-space to an L-space, (vii) and (viii) apply simultaneously.

18.6 Example. If T is any operator on ℓ_p^n, then $\pi^+(T) = (\Sigma \|Te_j\|^{p'})^{1/p'}$ (= K, say); in particular, $\pi^+(T) = \pi_2(T)$ for operators on ℓ_2^n. For if $x \geqslant 0$, then $Tx = \Sigma\, x(j)(Te_j)$, so $\|Tx\| \leqslant \Sigma \|Te_j\|\, x(j)$, which is of the form f(x), with $\|f\| = K$. Also, there exist $\lambda_j \geqslant 0$ such that $\Sigma\, \lambda_j^{\,p} = 1$ and $\Sigma\, \lambda_j \|Te_j\| = K$. Then $\|\Sigma\, \lambda_j e_j\| = 1$, while $\Sigma\, \|T(\lambda_j e_j)\| = K$.

18.7 Example. Define T on $\mathbb{R}^2$ by : $T(x,y) = (x+y, x-y)$. Then $\|T_{\infty,1}\| = 2$, while $\pi^+(T_{\infty,1}) = \|Te_1\|_1 + \|Te_2\|_1 = 4$. This shows the need for T to be positive in the second statement in (vii). Further, by (vii), $\pi^+(T_{1,\infty}) = \|T_{1,\infty}\| = 1$, while

$$\pi^+(T_{1,\infty}T_{\infty,1}) = 2\pi^+(I_{\infty,\infty}) = 4 .$$

This shows the need for A to be positive in (iv).

Exercise. Show that an operator T is cone-summing if and only if $\Sigma\, \|Tx_n\|$ is convergent for every Cauchy series $\Sigma\, x_n$ of positive elements.

Trace duality

By amending the definition of ν_∞ in the obvious way, we can easily identify the predual of π^+ under trace duality. As before, let X be a normed lattice and Y any normed linear space. Let S be an element of FL(Y,X). Define

$$\nu^+(S) = \inf\{\, \|\Sigma x_i\| : S = \Sigma\, g_i \otimes x_i \text{ with } \|g_i\| = 1,\ x_i \geqslant 0\},$$

in which, as always, all possible finite representations are considered. Such representations exist, since if $S = \Sigma\, g_i \otimes u_i$, where the u_i are not necessarily positive, then

$$S = \Sigma\, g_i \otimes u_i^+ + \Sigma\, (-g_i) \otimes u_i^- .$$

Since $u_i^+ + u_i^- = |u_i|$, this gives the alternative formulation

$$\nu^+(S) = \inf\{\, \|\Sigma\, |u_i|\, \| : S = \Sigma\, g_i \otimes u_i \text{ with } \|g_i\| = 1\} .$$

It is now clear that $\nu_\infty(S) \leq \nu^+(S) \leq \nu_1(S)$, with $\nu^+(S) = \nu_\infty(S)$ when X is an M-space and $\nu^+(S) = \nu_1(S)$ when X is an L-space.

18.8. Under trace duality, π^+ is the dual of ν^+. Further, if X,Y are both normed lattices and T is a positive cone-summing operator from X to Y, then

$$\pi^+(T) = \sup \{\text{trace } (TS) : S \in FL(Y,X),\ S \geq 0 \text{ and } \nu^+(S) \leq 1\} .$$

Proof. Let X be a normed lattice and T a cone-summing operator from X to Y. Suppose that $S \in FL(Y,X)$ is represented by $\Sigma\, g_i \otimes x_i$, with $\|g_i\| = 1$ and $x_i \geq 0$. Then

$$\text{trace } (TS) = \Sigma\, g_i(Tx_i) \leq \Sigma\, \|Tx_i\|$$
$$\leq \pi^+(T)\, \|\Sigma\, x_i\| .$$

Hence trace $(TS) \leq \pi^+(T)\nu^+(S)$.

Given $\varepsilon > 0$, there exist elements x_i of X_+ such that $\|\Sigma\, x_i\| = 1$ and $\Sigma\, \|Tx_i\| \geq (1-\varepsilon)\pi^+(T)$. There are elements g_i of Y^* (positive if Y is a lattice and $Tx_i \geq 0$) such that $\|g_i\| = 1$ and $g_i(Tx_i) = \|Tx_i\|$. The statement follows on putting $S = \Sigma\, g_i \otimes x_i$.

Exercise. If S is a positive operator on ℓ_∞^n, show that $\nu^+(S) = \|S\|$.

The "Pietsch" theorem

We now establish an analogue of Pietsch's theorem for cone-summing operators. It has two attractive special features. Firstly, it applies to X itself, without any embedding. Secondly, the proof (which is basically similar) delivers a linear function directly, instead of a superlinear one. Hence it is "constructive" instead of depending on the Hahn-Banach theorem. We need two algebraic lemmas, both very familiar in linear lattice theory.

18.9 Lemma. Let x,y and $z_1, \dots, z_k$ be positive elements of a linear lattice such that $x+y = z_1 + \dots + z_k$. Then there exist positive elements x_i, y_i such that $z_i = x_i + y_i$ for each i and $\Sigma\, x_i = x$, $\Sigma\, y_i = y$.

Proof. We show first that if $0 \leq u \leq x+y$, then u can be expressed as $x_0 + y_0$, where $0 \leq x_0 \leq x$ and $0 \leq y_0 \leq y$. Let $x_0 = x \wedge u$. Then $0 \leq x_0 \leq x$. The element u-y is less than both x and u, so $u-y \leq x_0$. Hence if $y_0 = u - x_0$, then $0 \leq y_0 \leq y$, as required.

We now prove the given statement by induction on k. It is trivial for $k = 1$. Assume it for a certain k, and suppose that $x+y = z_1 + \ldots + z_{k+1}$. By the above, z_{k+1} can be expressed as $x_{k+1} + y_{k+1}$, where $0 \leq x_{k+1} \leq x$ and $0 \leq y_{k+1} \leq y$. Write $x' = x - x_{k+1}$, $y' = y - y_{k+1}$. Then $x' + y' = z_1 + \ldots + z_k$, and the induction hypothesis gives the result.

<u>18.10 Lemma.</u> Let X be a linear lattice, and f a real-valued function on X_+ such that $f(x+y) = f(x) + f(y)$ and $f(\lambda x) = \lambda f(x)$ for all x,y in X_+ and $\lambda \geq 0$. Then f can be extended to a linear functional on X.

Proof. Define $f(x) = f(x^+) - f(x^-)$ for $x \in X$. This is consistent with the definition on X_+. We show that f is additive on X. Choose $x,y \in X$. We have

$$(x+y) = (x+y)^+ - (x+y)^- = (x^+ - x^-) + (y^+ - y^-) ,$$

so

$$(x+y)^+ + x^- + y^- = (x+y)^- + x^+ + y^+ .$$

By applying f to both sides and collecting terms, we obtain $f(x+y) = f(x) + f(y)$.

<u>18.11 Proposition.</u> Let T be a cone-summing operator on a normed lattice X. Then there is a positive functional f on X such that $\|f\| = \pi^+(T)$ and $\|Tx\| \leq f(|x|)$ for all $x \in X$.

Proof. For $x \geq 0$, define

$$f(x) = \sup \{ \Sigma \|Tx_i\| : x = \Sigma x_i \text{ with each } x_i \geq 0 \},$$

in which all finite positive decompositions of x are considered. From the definition, we have at once that $0 \leq f(x) \leq \pi^+(T)\|x\|$. Also, $f(x) \geq \|Tx\|$. Clearly $f(x+y) \geq f(x) + f(y)$ for x,y in X_+. We show that in fact equality holds. Suppose that $x+y = \Sigma z_i$, with $z_i \geq 0$. There are positive elements x_i, y_i as in 18.9. Then

$$\Sigma \|Tz_i\| \leq \Sigma \|Tx_i\| + \Sigma \|Ty_i\| \leq f(x) + f(y) ,$$

and hence $f(x+y) \leq f(x) + f(y)$, as required.

Extend f to X, as in 18.10. Then $\|f\| \leq \pi^+(T)$, by 18.2, and for any x in X,

$$\|Tx\| \leq \|Tx^+\| + \|Tx^-\| \leq f(x^+) + f(x^-) = f(|x|) .$$

Note that for an operator on an L-space, the required functional f is simply given by $f(x) = \|T\|.\|x\|$ for $x \geq 0$, and for a positive operator into an L-space, by $f(x) = \|Tx\|$ for $x \geq 0$.

We now derive factorization and extension theorems analogous to those for 2-summing operators, with the Hilbert space replaced by an L-space. Given a positive functional f, let $K = \{x : f(|x|) = 0\}$. If $x \in K$ and $|y| \leq |x|$, then $y \in K$ (that is, K is a "lattice ideal"). It follows that X/K is a linear lattice (we omit the details); it becomes a normed lattice with the norm $\|Qx\|_f = f(|x|)$, where Q is the quotient mapping. For positive x,y, we have

$$\|Qx + Qy\|_f = f(x+y) = \|Qx\|_f + \|Qy\|_f ,$$

so $(X/K, \| \|_f)$ is an L-space. By 18.5(vii), $\pi^+(Q) \leq \|f\|$. This construction, together with 18.11, gives:

18.12 Proposition. Let T be a cone-summing operator from a normed lattice X to a normed linear space Y. Then there exist an L-space L, a positive operator $T_1 : X \to L$ and an operator $T_2 : L \to Y$ (positive if T is positive) such that $T = T_2T_1$ and $\pi^+(T_1) = \|T_1\| = \pi^+(T)$, $\pi^+(T_2) = \|T_2\| = 1$.

Proof. With f as in 18.11, let L be the space X/K just constructed, and let $T_1 = Q$. If $Qx = 0$, then $f(|x|) = 0$, so $Tx = 0$. Hence it is consistent to define $T_2(Qx) = Tx$, and we have $\|T_2(Qx)\| \leq f(|x|) = \|Qx\|_f$, so $\|T_2\| \leq 1$. Since T_2 is defined on an L-space, $\|T_2\| = \pi^+(T_2)$.

If X or Y is complete (i.e. a Banach space), then we can take L to be complete.

The following extension theorem, due to Lotz (1975), is one of the reasons for the special position of L-spaces in the theory of Banach lattices. For a proof, we refer to Schaefer [BLPO] or Donner (1982).

Lotz's theorem. Let X be a linear sublattice of a normed lattice $\bar{X}$, and let L be a complete L-space. Let T be a positive operator from X to L. Then T has a positive extension $\bar{T} : \bar{X} \to L$ with $\|\bar{T}\| = \|T\|$.

Assuming this, we obtain at once the promised extension theorem for cone-summing operators:

18.13 Proposition. Let X be a linear sublattice of a normed lattice $\bar{X}$, and let Y be a Banach space. Let T be a cone-summing operator from X to Y. Then T has an extension $\bar{T} : \bar{X} \to Y$ with $\pi^+(\bar{T}) = \pi^+(T)$. If Y is also a normed lattice and T is positive, then the extension may be assumed to be positive.

Proof. Express T as T_2T_1 as in 18.12, and apply Lotz's theorem to T_1.

In particular, a closed sublattice of an L-space admits a positive projection of norm 1. The existence of non-complemented subspaces of ℓ_1 (or badly complemented subspaces of ℓ_1^n) is enough to show that we really need the condition that X is a sublattice of $\bar{X}$ (not just a subspace).

Note that for operators on an M-space, this amounts to a theorem on extension with preservation of π_1.

Exercise. Prove Lotz's theorem for operators into ℓ_1^n by equating such operators with functionals on X^n and using the sublinear function on X^n defined by $q(x_1, \dots, x_n) = \|x_1^+ \vee \dots \vee x_n^+\|$.

Finally, we describe a dual characterization of π^+. Let X be a normed lattice and S an operator from Y to X. The majorizing norm of S is

$$\alpha(S) = \sup \{ \| |Sy_1| \vee \dots \vee |Sy_n| \| : y_i \in U_Y, n \in \mathbb{N} \}.$$

18.14. Let T be an operator from a normed lattice X to a normed linear space Y. Then $\pi^+(T) = \alpha(T^*)$.

Proof. Choose positive elements x_i with $\| \Sigma x_i \| \leq 1$. For each i, let $g_i \in Y^*$ be such that $\|g_i\| = 1$ and $g_i(Tx_i) = \|Tx_i\|$. Let $h = (T^*g_1) \vee \dots \vee (T^*g_n)$. Then $\|h\| \leq \alpha(T^*)$, and

$$\sum_i \|Tx_i\| = \sum_i (T^*g_i)(x_i) \leq \sum_i h(x_i) \leq \|h\| .$$

Hence $\pi^+(T) \leq \alpha(T^*)$.

To prove the reverse inequality, let f be the functional given by 18.11. Given any $g \in U_{Y^*}$ and $x \in X_+$, we have $\pm\, g(Tx) \leq \|Tx\| \leq f(x)$, so that $|T^*g| \leq f$. It now follows at once that $\alpha(T^*) \leq \|f\| = \pi^+(T)$.

<u>Exercise.</u> For operators S from Y to X, show that $\pi^+(S^*) = \alpha(S)$.

For further results on cone-summing norms (and for a systematic account of the theory of normed lattices), we refer to [BLPO].

REFERENCES

Books for further reading

[BLPO] Schaefer, H.H. Banach Lattices and Positive Operators. Berlin: Springer (1974).

[BMOI] Tomczak-Jaegermann, N. Banach-Mazur Distances and Finite Dimensional Operator Ideals. Boston : Pitman (to appear).

[CBS I,II] Lindenstrauss, J. & Tzafriri, L. Classical Banach Spaces I,II. Berlin: Springer (1977, 1979).

[FDBS] Pelczynski, A. Geometry of Finite Dimensional Banach Spaces. In Notes in Banach Spaces, ed. H.E. Lacey. Austin, Texas : University of Texas (1980).

[FLO] G. Pisier. Factorization of Linear Operators and Geometry of Banach Spaces. Providence, Rhode Island : American Mathematical Society CBMS 60 (1986).

[OI] Pietsch, A. Operator Ideals. Amsterdam : North Holland (1980).

Other works referred to in the text

Donner, K. (1982). Extension of Positive Operators and Korovkin Theorems. Berlin : Springer Lecture Notes 904.

Dubinsky, E., Pelczysnki, A. & Rosenthal, H.P. (1972). On Banach spaces X for which $\Pi_2(\mathcal{L}_\infty,X) = B(\mathcal{L}_\infty,X)$. Studia Math. 44, 617-648.

Dvoretzky, A. & Rogers, C.A. (1950). Absolute and unconditional convergence in normed linear spaces. Proc. Nat. Acad. Sci. U.S.A. 36, 192-197.

Garling, D.J.H. & Gordon, Y. (1971). Relations between some constants associated with finite dimensional Banach spaces. Israel J. Math. 9, 346-361.

Gluskin, E.D. (1981). The diameter of the Minkowski compactum is roughly equal to n (Russian). Funk. Anal. i Prilozh. 15, 72-73.

Gordon, Y. (1968). On the projection and Macphail constants of ℓ_n^p spaces. Israel J. Math. 6, 295-302.

Gordon, Y. (1969). On p-absolutely summing constants of Banach spaces. Israel J. Math. 7, 151-163.

Gordon, Y. & Lewis, D.R. (1974). Absolutely summing operators and local unconditional structures. Acta Math. 133, 27-48.

Grothendieck, A. (1955). Produits tensoriels topologiques et espaces nucleaires. Mem. Amer. Math. Soc. 16.

Grothendieck, A. (1956). Resume de la theorie metrique des produits tensoriels topologiques. Bol. Soc. Mat. Sao Paulo 8, 1-79.

Grünbaum, B. (1960). Projection constants. Trans. Amer. Math. Soc. 95, 451-465.
Haagerup, U. (1982). The best constants in the Khintchine inequality. Studia Math. 70, 231-283.
Haagerup, U. (1985). The Grothendieck inequality for bilinear forms on C*-algebras. Adv. in Math. 56, 93-116.
Haagerup, U. (1987). An upper bound for the complex Grothendieck constant. Israel J. Math., to appear.
Holub, J.R. (1970). Tensor product mappings. Math. Ann. 188, 1-12.
Jameson, G.J.O. (1987). Relations between summing norms of mappings on ℓ_∞^n. Math. Z. 194, 89-94.
John, F. (1948). Extremum problems with inequalities as subsidiary conditions. In Courant Anniversary Volume. New York : Interscience.
Johnson, W.B., Rosenthal, H.P. & Zippin, M. (1971). On bases, finite dimensional decompositions and weaker structures in Banach spaces. Israel J. Math. 9, 488-506.
Kadec, M.I. & Snobar, M.G. (1971). Certain functionals on the Minkowski compactum (Russian). Mat. Zametki 10, 453-458.
Khinchin, A. (1923). Uber dyadische Brüche. Math. Z. 18, 109-116.
König, H. (1985). Spaces with large projection constants. Israel J. Math. 50, 181-188.
Krivine, J.L. (1973-74). Theoremes de factorisation dans les espaces reticules. Seminaire Maurey-Schwarz, exp. 22-23.
Krivine, J.L. (1979). Constantes de Grothendieck et fonctions de type positif sur les spheres. Adv. in Math. 31, 16-30.
Kwapien, S. (1972). Isomorphic characterizations of inner product spaces by orthogonal series with vector coefficients. Studia Math. 44, 583-595.
Lindenstrauss, J. & Pelczynski, A. (1968). Absolutely summing operators in $\mathcal{L}_p$ spaces and their applications. Studia Math. 29, 275-326.
Lotz, H.P. (1975). Extensions and liftings of positive linear mappings on Banach lattices. Trans. Amer. Math. Soc. 211, 85-100.
Macphail, M.S. (1947). Absolute and unconditional convergence. Bull. Amer. Math. Soc. 53, 121-123.
Maurey, B. (1974a). Theoremes de factorisation pour les operateurs lineaires a valeurs dans les espaces L^p. Soc. Math. de France, Asterisque 11.
Maurey, B. (1974b). Type et cotype dans les espaces munis de structures locales inconditionelles. Seminaire Maurey-Schwartz 1973-74, no. 24-25.
Maurey, B. & Pisier, G. (1976). Series de variables aleatoires vectoriels independantes et proprietes geometriques des espaces de Banach. Studia Math. 58, 45-90.
Orlicz, W. (1933). Uber unbedingte Konvergenz in Funktionenräumen I. Studia Math. 4, 33-37.
Pelczynski, A. (1962). Proof of Grothendieck's theorem on the characterization of nuclear spaces (Russian). Prace Mat. 7, 155-167.
Persson, A. & Pietsch, A. (1969). p-nukleare und p-integrale Abbildungen in Banachräumen. Studia Math. 33, 19-62.
Pietsch, A. (1961). Unbedingte und absolute Summierbarkeit in F-Räumen. Math. Nachrichten 23, 215-222.
Pietsch, A. (1963). Absolut summierende Abbildungen in lokalkonvexen Räumen. Math. Nachrichten 27, 77-103.
Pietsch, A. (1967). Absolut p-summierende Abbildungen in normierten Räumen. Studia Math. 28, 333-353.
Pisier, G. (1978). Grothendieck's theorem for non-commutative C*-algebras, with an appendix on Grothendieck's constants. J. Funct. Anal. 29, 397-415.

Pisier, G. (1986). Factorization of operators through $L_{p\infty}$ or L_{p1} and non commutative generalisations. Math. Ann. 276, 105-136.

Rosenthal, H.P. (1976). Some applications of p-summing operators to Banach Banach space theory. Studia Math. 58, 21-43.

Schlotterbeck, U. (1971). Uber Klassen majorisierbarer Operatoren auf Banachverbänden. Rev. Acad. Ci. Zaragoza (2) 26, 585-614.

Schütt, C. (1978). Unconditionality in tensor products. Israel J. Math. 31, 209-216.

Szarek, S. (1976). On the best constants in the Khinchine inequality. Studia Math. 58, 197-208.

Tomczak-Jaegermann, N. (1974). The moduli of smoothness and convexity and the Rademacher averages of the trace classes S_p ($1 \leq p < \infty$). Studia Math. 50, 163-182.

Tomczak-Jaegermann, N. (1979). Computing 2-summing norm with few vectors. Arkiv för Mat. 17, 273-277.

LIST OF SYMBOLS

Symbol	*name (if any)*	*page*
d	Banach-Mazur distance	2
λ	projection constant	7
ν_1	1-nuclear norm	17
μ_p		24
π_p	p-summing norm	31
$\pi_p^{(n)}$		33
$\pi_{q,p}$	(q,p)-summing norm	42
ν_p	p-nuclear norm	46
D_n	$\{-1,1\}^n$	64
ρ_2		69
τ_2	type 2 constant	69
κ_2	cotype 2 constant	69
Δ_2	2-dominated constant	97
β	basis constant	123
i_p	p-integral norm	151
π^+	cone-summing norm	163

INDEX